Student's Guide and Solutions Manual

Bernard Gillett
University of Colorado at Boulder

Using and Understanding Mathematics
A Quantitative Reasoning Approach

Fourth Edition

Jeffery O. Bennett
University of Colorado at Boulder

William L. Briggs
University of Colorado at Denver

PEARSON

Addison
Wesley

Boston San Francisco New York
London Toronto Sydney Tokyo Singapore Madrid
Mexico City Munich Paris Cape Town Hong Kong Montreal

Reproduced by Pearson Addison-Wesley from electronic files supplied by the author.

Copyright © 2008 Pearson Education, Inc.
Publishing as Pearson Addison-Wesley, 75 Arlington Street, Boston, MA 02116.

ISBN-13: 978-0-321-46022-6
ISBN-10: 0-321-46022-7

3 4 5 6 BRR 10 09 08

Table of Contents

Study Guide

Solutions

Introduction

Welcome to the *Study Guide and Solutions Manual* for *Using and Understanding Mathematics*. Hopefully, with the help of this book, your course in quantitative reasoning will be both enjoyable and successful. The goal of this guide is not to add to your workload, but to give you a set of concise notes that will make your studying as effective as possible. If you work with this guide as you read and do exercises, you should get the most from the course.

This guide is organized according to the units in the textbook. For each unit you will find the following features:

Overview

This section provides a brief survey of the unit, its major points and its goals. This feature is not a substitute for reading the text.

Key Words and Phrases

This section is simply a list of the important words and phrases used in the unit. You can use this section for review, study, and self-testing. You should be able to explain or define all of the terms on this list.

Key Concepts and Skills

In this section you will find a summary of the most important concepts in each unit. These concepts may be general ideas (for example, the distinction between deductive and inductive arguments) or basic skills (for example, creating the equation of a straight line). This section should also be helpful for review, study, and self-testing.

Important Review Boxes

If a unit has one or more Review Boxes, they are listed. These boxes are quite important for providing background skills or knowledge needed for the unit.

Solutions to (Most) Odd Problems

Following all of the unit summaries, the second half of this Guide provides full solutions to most of the odd-numbered problems in the textbook. Mathematics is not a spectator sport. Reading the solutions is never a substitute for working the problems. *You are strongly advised to work the problems first and then check the solutions.*

How to Succeed In This Course

Using *This* Book

Before we get into more general strategies for studying, here are a few guidelines that will help you use *this* book most effectively.

- Before doing any assigned problems, read assigned material *twice*:
    On the first pass, read quickly to gain a feel for the material and concepts presented.
    On the second pass, read the material in more depth, and work through the examples carefully.
- During the second reading, take notes that will help you when you go back to study later. In particular, *use the margins.* The wide margins in this textbook are designed to give you plenty of room for making notes as you study.
- After you complete the reading, and again when studying for exams, make sure you can answer the *review questions* at the end of each unit.
- You'll learn best by *doing*, so work plenty of the end-of-unit exercises. Don't be reluctant to work more than the exercises that your instructor assigns.

Budgeting Your Time

A general rule of thumb for college courses is that you should expect to study about 2 to 3 hours per week *outside* class for each unit of credit. For example, a student taking 15 credit hours should spend 30 to 45 hours each week studying outside of class. Combined with time in class, this works out to a total of 45 to 60 hours per week — not much more than the time required of a typical job. If you find that you are spending fewer hours than these guidelines suggest, you can probably improve your grade by studying more. If you are spending more hours than these guidelines suggest, you may be studying inefficiently; in that case, you should talk to your instructor about how to study more effectively for a mathematics class.

General Strategies for Studying

- Don't miss class. Listening to lectures and participating in discussions is much more effective than reading someone else's notes. Active participation will help you retain what you are learning.
- Budget your time effectively. An hour or two each day is more effective, and far less painful, than studying all night before homework is due or before exams.
- If a concept gives you trouble, do additional reading or problem solving beyond what has been assigned. If you still have trouble, *ask for help*: you surely can find friends, colleagues, or teachers who will be glad to help you learn. Never be reluctant to ask questions or ask for help in this course. If you have a question or problem, it is extremely unlikely that you will be alone.

- Working together with friends can be valuable; you improve your own understanding when discussing concepts with others. However, be sure that you learn *with* your friends and do not become dependent on them.

Preparing for Exams

- Rework exercises and other assignments; try additional exercises to be sure you understand the concepts. Study your assignments, quizzes, and exams from earlier in the semester.
- Study your notes from lectures and discussions. Pay attention to what your instructor expects you to know for an exam.
- Reread the relevant sections in the textbook, paying special attention to notes you have made in the margins.
- Study individually before joining a study group with friends. Study groups are effective only if *every* individual comes prepared to contribute.
- Try to relax before and during the exam. If you have studied effectively, you are capable of doing well. Staying relaxed will help you think clearly.

Finally, good luck. We wish you an enjoyable and rewarding experience with quantitative reasoning.

1   Thinking Critically

Overview

Before discussing Chapter 1, we urge you to take a few minutes to read the prologue to the textbook. This short chapter sets the stage for the entire book. It presents the idea of quantitative reasoning and discusses the importance of interdisciplinary thinking. It gives a high altitude picture of mathematics and how it impacts many other subjects that you will encounter either in other courses or in your career. Finally, it gives some advice on using the book and studying for your course. It's worth a quick reading.

In teaching this course to many students of over many years, we know that often the most serious weakness that students bring to the course is not poor mathematical skills, but poor reasoning skills. It's not multiplying two numbers that creates problems, but deciding *when* to multiply. For this reason, the book opens with a chapter that contains virtually no mathematics (in the sense of manipulating equations and finding their solutions). The emphasis of the chapter is critical thinking and logical skills.

In this chapter you will encounter some introductory logic, but don't worry; we don't get carried away with symbolic logic and heavy-duty truth tables. In fact, much of this chapter may be familiar to you from a previous course in logic or philosophy.

Unit 1A Recognizing Fallacies

Unit 1A opens the chapter by explaining that a logical **argument** (as opposed to an everyday argument) is a set of facts or assumptions, called **premises**, that lead to a **conclusion**. A **fallacy** is an argument that is either deceptive or wrong. This unit explores common fallacies that you might encounter in advertising or news reports. We present ten informal fallacies, some of which may seem obvious, others of which may be quite subtle. Critical reading and thinking will help you avoid becoming a victim of these fallacies.

Key Words and Phrases

logic	argument	premise
conclusion	fallacy	appeal to popularity
false cause	appeal to ignorance	hasty generalization
limited choice	appeal to emotion	personal attack
circular reasoning	diversion	straw man

Key Concepts and Skills

- identify the premise and conclusion of an argument.
- recognize informal fallacies in advertisements and news reports.

Unit 1B Propositions and Truth Values

Overview

In this unit formal logic is introduced in a somewhat casual manner. We start with **propositions** — statements that make a claim that can be true or false. Then we look at the **connectors** that can be used with propositions to make more complex propositions. The connectors that you will encounter are

- *not* (negation)
- *or* (disjunction)
- *and* (conjunction)
- *if ... then* (implications).

Whereas many logic books make heavy use of symbolic logic and truth tables, we will use truth tables primarily for fairly simple propositions that involve one, two, or three connectors. So our excursion into symbolic logic will be limited and designed to provide only an introductory glimpse.

The *if ... then* connector is quite important in both logic and everyday speech (for example, *if I pass this course, then I will graduate*). For this reason, we spend a little time discussing other forms of the proposition *if p, then q*. These other forms are the

- **converse** (*if q, then p*)
- **inverse** (*if not p, then not q*) and
- **contrapositive** (*if not q, then not p*).

This particular discussion may seem a bit technical, but it's also extremely practical. For example, suppose it's true that *if I read the book, then I will pass the course*. Does it follow that *if I don't read the book, then I won't pass the course*?

Key Words and Phrases

proposition	truth table	negation
conjunction	disjunction	conditional
antecedent	consequent	converse
inverse	contrapositive	logical equivalence

Key Concepts and Skills

- understand negation, conjunction, disjunction, and conditionals, and their truth tables.

- use truth tables to evaluate the truth of compound propositions that use two or more connectors.
- analyze various forms of *if … then* propositions.

Unit 1C Sets and Venn Diagrams

Overview

You may have encountered Venn diagrams before now as a way to illustrate the relationships between collections of objects, or **sets**. In this unit, we review some of the most basic properties of sets and then discuss how Venn diagrams can be used to work with sets. Throughout the unit, the emphasis is on practical applications of sets and Venn diagrams. One of the important applications of Venn diagrams is to illustrate what are called **categorical propositions** of logic. We will study four basic categorical propositions. Given a **subject set** S and a **predicate set** P, they may be related in the following ways:

- All S are P (for example, all whales are mammals)
- No S are P (for example, no fish is a mammal)
- Some S are P (for example, some doctors are women)
- Some S are not P (for example, some teachers are not men).

As we will see, each form of categorical proposition has a specific Venn diagram. Equally important, the negation of each categorical proposition is one of the other propositions in the list above. Specifically, we have the following relations between the four propositions and their negations.

Proposition	Negation
All S are P	Some S are not P
No S are P	Some S are P
Some S are P	No S are P
Some S are not P	All S are P

The unit concludes with a discussion of the additional uses of Venn diagrams.

Key Words and Phrases

set	Venn diagram	subset
disjoint sets	overlapping sets	categorical propositions

Key Concepts and Skills

- use set notation.
- construct Venn diagrams for categorical propositions.
- put propositions in standard form.

- negate categorical propositions.
- construct Venn diagrams for three or more sets.

Important Review Box

- A Brief Review of Sets of Numbers

Unit 1D Analyzing Arguments

Overview

The propositions that we studied in the previous unit can be combined in various ways to form arguments. Of primary importance is the distinction between **deductive** and **inductive** arguments. Deductive arguments generally proceed from general premises to a more specific conclusion. In a deductive argument, all of the premises are needed to reach the conclusion. By contrast, inductive arguments generally proceed from specific premises to a general conclusion; in an inductive argument the premises independently support the conclusion.

To analyze inductive arguments, we ask about the **strength** or **weakness** of the argument. Determining the strength of an inductive argument is often a subjective judgment, and so there are no systematic methods to apply.

Most of the unit is spent analyzing three-line deductive arguments using Venn diagrams. They can consist of the four types of categorical propositions or they may involve conditional propositions (studied in Unit 1C). Of fundamental importance in this business is the distinction between valid and invalid arguments. An argument is **valid** if, based on the Venn diagram analysis, it is logically solid and consistent. An argument that fails the Venn diagram analysis must contain a fallacy and is **invalid**. Validity has nothing to do with the truth of the premises or conclusion, rather, it is a measure of the logical structure of the argument.

Having shown that an argument is valid, we can then ask if it is sound. A **sound** argument is valid *and* has true premises; a sound argument must lead to a true conclusion. Soundness is the highest test of a deductive argument.

Fallacies can arise in deductive arguments in many ways. Perhaps the most common fallacies occur in arguments that involve conditional (*if...then*) propositions. These fallacies appear in the everyday arguments of advertising and news reports. There are four different forms of conditional arguments; two are valid and two are invalid:

- affirming the hypothesis (valid)
- affirming the conclusion (invalid)
- denying the hypothesis (invalid)
- denying the conclusion (valid).

The unit closes with examples of how inductive and deductive arguments are used in mathematics.

Key Words and Phrases

deductive	inductive	strength/weakness
valid/invalid	sound	affirming the hypothesis
affirming the conclusion	denying the hypothesis	denying the conclusion
Pythagorean theorem		

Key Concepts and Skills

- know the distinction between deductive and inductive arguments.
- determine the strength of inductive arguments.
- assess the validity of three-line deductive arguments consisting of categorical propositions, using Venn diagrams.
- assess the validity of three-line deductive arguments involving conditional propositions, using Venn diagrams.
- identify fallacies that arise in conditional arguments.
- understand various combinations of valid/invalid and sound/unsound that can occur in deductive arguments.
- determine the soundness of three-line deductive arguments.

Unit 1E Critical Thinking in Everyday Life

Overview

Critical thinking is an approach to problem solving and decision-making that involves careful reading (or listening), sharp thinking, logical analysis, good visualization, and healthy skepticism. In this chapter, we present several guidelines designed to sharpen critical thinking skills as they apply to practical problems:

- Read or listen carefully.
- Look for hidden assumptions.
- Identify the real issue.
- Use visual aids (pictures, diagrams, tables).
- Understand all the options.
- Watch for fine print and misinformation.
- Are other conclusions possible?

Key Concepts and Skills

- apply the guidelines of the unit to practical decisions and problems.

2   Approaches to Problem Solving

Overview

The first chapter of the book was devoted to *qualitative* issues — topics that don't require extensive use of numbers and computation. In this chapter (and the remainder of the book) we turn to *quantitative* matters. Perhaps it's not surprising that we begin our study of quantitative topics with problem solving. The first two units of the chapter deal with a very basic and important problem solving technique, the use of units. The last unit of the chapter presents various problem-solving strategies. It is a valuable chapter whose lessons run through the rest of the book.

Unit 2A The Problem-Solving Power of Units

Overview

Nearly every number that you encounter in the real world is a measure of *something*: 6 billion *people*, 5280 *feet*, 9 trillion *dollars*, 26 *cubic feet*. The quantities that go with numbers are called **units**. The message of this section of the book is that using units can simplify problem solving immensely. Indeed the use of units is one of the most basic problem solving tools.

We begin by considering the most basic **simple units** for the fundamental types of measurement. Here are a few examples of simple units.
- length — inches, feet, meters.
- weight — pounds, grams.
- volume — quarts, gallons, liters.
- time — seconds, hours.

From these simple units, we can build endless **compound units**.

The rest of the unit is spent illustrating a wide variety of such compound units. Among the many you will meet and work with are…
- units of area such as square feet and square yards.
- units of volume such as cubic inches and cubic feet.
- units of speed such as miles per hour.
- units of price such as dollars per pound.
- units of gas mileage such as miles per gallon.

One of the realities of life is that there are many units for the same quantity. For example, we can measure lengths in inches, centimeters, feet, meters, miles, or kilometers. We measure floor area in square feet, but have to buy carpet in square yards. Thus one of the necessities of problem solving is being able to convert from one unit to another consistent unit. The key to carrying out conversions between units is to realize that there are three equivalent ways to express a conversion factor. For example, we can say 1 foot = 12 inches, or we can say 1 foot *per* 12 inches, or we can say 12 inches *per* foot. Mathematically, we can write

$$1 \text{ ft} = 12 \text{ in. or } \frac{1 \text{ ft}}{12 \text{ in.}} = 1 \text{ or } \frac{12 \text{ in.}}{1 \text{ ft}} = 1.$$

These three forms of the same conversion factor are equivalent. The key to a happy life with units is choosing the appropriate form of the conversion factor for a given situation.

Additional problem solving skills can then be built on these fundamental skills. You will see and learn how to make a chain of conversions factors to solve more complex problems. For example, the time required to count a billion dollars at the rate of one dollar per second is

$$\$1,000,000,000 \times \frac{1 \text{ sec}}{\$1} \times \frac{1 \text{ min}}{60 \text{ sec}} \times \frac{1 \text{ hr}}{60 \text{ min}} \times \frac{1 \text{ day}}{24 \text{ hr}} \times \frac{1 \text{ yr}}{365 \text{ days}} = 31.7 \text{ years.}$$

The fact that the units cancel and give an answer in *years* tells you that the problem has been set up correctly.

Finally, we look at a very practical type of unit conversion problem, those associated with currency. If you travel to France you might find 1 franc is equal to about 18 cents. From this fact, you might need to answer questions such as

- which is larger, 1 franc or 1 dollar?
- how many francs in a dollar?
- how many dollars in a franc?
- how many dollars in a 23.45 francs?
- if apples cost 23 francs per kilogram, what is the price in dollars per pound?

Here are a few final words of advice: Never was the motto *practice makes perfect* more true than with problem solving and units. You should work all assigned exercises, and then some, in order to master these techniques. And unless your instructor tells you otherwise, there is no need to memorize hundreds of conversion factors. It's helpful to know a few essential conversion factors off the top of your head. As for the rest, it's easiest just to know how to find them quickly in the book.

Key Words and Phrases

simple units	compound units	area
volume	conversion factor	

Key Concepts and Skills

- convert from one simple unit to another; for example, from inches to yards.

- convert from one unit of area to another unit of area; for example, from square inches to square yards.
- convert from one unit of volume to another unit of area; for example, from cubic inches to cubic yards.
- solve problems involving chains of conversion factors; for example, finding the number of seconds in a year.
- convert from one unit of currency to another; for example, from dollars to pesos.

Important Review Box

- A Brief Review of Working with Fractions

Unit 2B Standardized Units: More Problem-Solving Power

Overview

In this unit we continue the study of units and conversion factors, but now explore the two standard systems of units: The U.S. Customary System of Measurement (or USCS system, which is used primarily in the Unites States) and the metric system (which is used everywhere else in the world).

We first proceed systematically and survey the USCS units for length, weight, and volume. Tables 2.2, 2.3, and 2.4 contain many conversion factors, but you should focus on *using* these conversion factors, not memorizing them. Having seen the complications of the USCS system, the metric system should come as a welcome relief. Next we present the metric units for length, weight and volume. The system is based on powers of ten and standard prefixes (Table 2.5), which makes conversions between units relatively simple.

Unfortunately, for people living in the United States, conversions between the metric system and the USCS system are necessary. Table 2.6 has a few of the essential conversion factors between the metric and USCS systems. You may want to write other useful conversion factors in the margin near this table.

Finally, we explore two more categories of units. **Energy** (what makes things move or heat up) and **power** (the rate at which energy is used) are incredibly important concepts in understanding the world around us. Uses of energy and power units are presented in matters such as utility bills, diet, and the environment. The units of **density** and **concentration** are also discussed in detail. In measuring population density, the capacity of computer discs, levels of pollution, and blood alcohol content, these units are indispensable.

Armed with all of these units and conversion factors (and don't forget currency conversion factors as well), we can do even more elaborate problem solving. This is the goal of the remainder of the unit and the exercises at the end of the unit.

Key Words and Phrases

USCS system	metric system	meter
gram	liter	second
metric prefixes	Celsius	Fahrenheit
Kelvin	energy	power
kilowatt-hour	watt	density
concentration	population density	

Key Concepts and Skills

- given the required conversion factor, convert between two consistent USCS units; for example, rods to miles.
- multiply and divide powers of ten.
- know basic metric units and commonly used prefixes.
- given the required conversion factor, convert between two consistent metric units; for example, millimeters to kilometers.
- given the required conversion factor, convert between a USCS unit and a consistent metric unit; for example, cubic inches to liters.
- convert between temperatures in the three standard systems.
- solve problems using chains of conversion factors involving USCS units, metric units, and currency units.
- understand and apply units of energy and power.
- understand and apply units of density to materials, population, and information.
- understand and apply units of concentration to pollution, and blood alcohol content.

Important Review Box

- A Brief Review of Powers of Ten

Unit 2C Problem Solving Guidelines and Hints

Overview

Having seen very specific examples of problem solving in the previous two units, we now present some strategies for problem solving in general. There is no simple and universal formula for solving all problems. Only continual (and hopefully enjoyable) practice can lead one towards mastery in problem solving. This unit is designed to provide you some of that practice.

The unit opens with a very well known four-step process for approaching problem solving. This process is not a magic formula, but rather a set of guidelines. They four basic steps are:

- understand the problem,
- devise a strategy,
- carry out the strategy, and
- look back, check, interpret and explain your solution.

The remainder of the unit is a list of eight strategic hints for problem solving. Here are the strategic hints:

- there may be more than one answer.
- there may be more than one strategy.
- use appropriate tools.
- consider simpler, similar problems.
- consider equivalent problems with simpler solutions.
- approximations can be useful.
- try alternative patterns of thought.
- don't spin your wheels.

Most of the unit consists of examples in which the four-step process and these strategic hints are put to use. Study these problems and solutions, and try to see how the techniques might be used in other problems that you encounter.

Key Concepts and Skills

- carry out the four-step problem solving process.
- identify and use the eight strategic problem solving hints.

---

3   Numbers in the Real World

Overview

In this chapter we explore how numbers are used in real and relevant problems. The first unit deals with percentages and is arguably one of the most practical and important units of the book. As you know, many real world numbers are incredibly large (the federal debt or the storage capacity of a computer disk) or very small (the diameter of a cancer cell or the wavelength of an x-ray), so in the next unit, we introduce scientific notation to deal with large and small numbers. Unit 3C discusses another reality of numbers in the real world: they are often approximate or subject to errors. The news is filled with reference to index numbers; for example, the consumer price index, or CPI. Such index numbers are critical to an understanding of the news, so they are the subject of Unit 3D. Finally, the last unit is a fascinating exploration of how numbers deceive us. All in all, it's a very practical and useful chapter.

Unit 3A Uses and Abuses of Percentages

Overview

If you read a news article, a financial statement, or an economic report, you may notice that one of the most common ways to communicate quantitative information is with percentages. In this unit, we will explore the many ways that percentages are used — and abused. Percentages are used for three basic purposes:

- as fractions (for example, 45% of the voters favored the incumbent),
- to describe change (for example, taxes increased by 10%), and
- for comparison (for example, women live 3% longer than men).

In this unit, we examine each of these uses of percentages in considerable detail with plenty of examples taken from the news and from real situations. The use of percentages as fractions is probably familiar. The use of percentages to describe change relies on the notions of **absolute change** and **relative** (or **percentage**) **change**. When some quantity changes from a previous value to a new value, we can define its change in either of two ways:

$$\text{absolute change} = \text{new value} - \text{reference value}$$

$$\text{relative change} = \frac{\text{absolute change}}{\text{reference value}} = \frac{\text{new value} - \text{reference value}}{\text{reference value}}.$$

The use of percentages for comparison relies on the notions of **absolute difference** and **relative** (or **percentage**) **difference**. To make comparisons between two quantities, we must identify the **compared quantity** (the quantity that we are comparing) and the **reference quantity** (the quantity to which we make the comparison). For example, if we ask how much larger is one meter than one yard, *one meter* is the compared quantity and *one yard* is the reference quantity. Then we can calculate two kinds of difference:

$$\text{absolute difference} = \text{compared value} - \text{reference value}$$

$$\text{relative difference} = \frac{\text{absolute difference}}{\text{reference value}} = \frac{\text{compared value} - \text{reference value}}{\text{reference value}}.$$

A powerful rule for interpreting statements involving percentages is what we call the **of versus more than rule**. It says that if the compared value is *P%* *more than* the reference value, then it is $(100 + P)\%$ *of* the reference value, and if the compared value is *P% less than* the reference value, it is $(100 - P)\%$ *of* the reference value. For example, if Bess' salary is 30% more than Bob's salary, then Bess' salary is 130% *of* Bob's salary. Similarly, if Jill's height is 30% *less than* Jack's height, then Jill's height is 70% *of* Jack's height.

All of these ideas are assembled in this unit to solve a variety of practical problems. Our experience in teaching this subject is that the greatest difficulty is understanding the problem and translating it into mathematical terms. It is important to read the problem carefully, draw a picture if necessary, decide how percentages are used in the problem (as a fraction, for change, or for comparison), and to write a mathematical sentence that describes the situation. It is important to study the examples in the unit and work plenty of practice problems.

The unit closes with several examples of ways that percentages are abused. Be sure you can identify problems in which the previous value shifts and problems in which the quantity of interest is itself a percentage.

Key Words and Phrases

absolute change	relative change	percentage change
reference quantity	compared quantity	absolute difference
relative difference	percentage difference	*of* versus *more than* rule

Key Concepts and Skills

- identity and solve problems that use percentages as fractions.
- identity and solve problems that use percentages to describe change.
- identity and solve problems that use percentages for comparison.
- apply the *of* versus *more than* rule to practical problems.
- identify fallacies and errors in statements involving percentages.

Important Review Boxes

- A Brief Review of Percentages
- A Brief Review of What is a Ratio?

Unit 3B Putting Numbers in Perspective

Overview

Numbers in the world around us are often very large or very small. In order to write large and small numbers compactly, without using a lot of zeros, we use **scientific notation**. This is the main new mathematical idea in this unit. If you haven't used scientific notation before, you will want to practice writing numbers in scientific notation, and multiplying and dividing numbers in scientific notation.

We next spend some time showing how to make rough calculations using estimation. Often we can use approximate values of quantities and arrive at useful **order of magnitude** answers.

All of us suffer from number numbness: large and small numbers eventually lose all meaning. For example, most of us have no sense of how large $9 trillion (the federal debt) or 6 billion (the world's population) really are. Therefore, just as important as *writing* large and small numbers is *visualizing* large and small numbers. We look at some fascinating methods, such as **scaling**, for giving large and small numbers meaning. The goal is often to associate numbers with a striking visual image. For example, if the earth is the size of a ball-point pen tip, the sun would be a grapefruit 15 meters away.

Key Words and Phrases

> scientific notation   order of magnitude   scaling
>
> scale ratio   light-year

Key Concepts and Skills

- write numbers in scientific notation.
- multiply and divide numbers in scientific notation.
- use estimates to compute order of magnitude answers.
- use scaling methods for visualizing large and small numbers.

Important Review Boxes

- A Brief Review of Working with Scientific Notation

Unit 3C Dealing With Uncertainty

Overview

While they may appear totally reliable, the numbers we see in the news or in reports often are quite uncertain and prone to errors. This unit reminds us of this fact in many different ways, as we consider the sources of errors and the measurement of uncertainty in numbers.

The precision of a number can be described by the number of **significant digits** it has. Significant digits are those digits in a number that we can assume to be reliable, although, this is where care must be used in reading numbers. It's usually easy to determine the number of significant digits in a number; there are a few subtle cases that are listed in the summary box in the unit.

Errors enter problems and calculation in two ways. Errors that arise in unpredictable and unavoidable ways are called **random errors**. On the other hand, **systematic errors** arise due to a problem in the system that affects all measurements in the same way; these errors can often be avoided.

We next discuss **absolute errors** and **relative errors**. It is important to note the analogy between absolute/relative errors and absolute/relative difference (discussed in Unit 3A).

The next observation in the unit concerns the difference between **accuracy** and **precision**. A pharmacist's scale that can weigh fractions of an ounce has much more *precision* than a bathroom scale that can measure only in pounds. Thus precision refers to how precisely a quantity can be measured or how precisely a number is reported.

On the other hand, the *accuracy* of a measurement refers to how close it is to the true value (which we often don't know). Suppose a marble weighs 4.5

ounces. If one scale reports a weight of 4.4 ounces and another scale reports a weight of 4.7 ounces, the first measurement is more accurate; both measurements have the same precision — to the nearest tenth of an ounce.

A few technicalities arise when it comes to doing arithmetic with approximate numbers. There are two rules that tell us how much precision should be assigned to the sum, difference, product, or quotient of two approximate numbers.

- when adding or subtracting two approximate numbers, the result should be rounded to the same precision as the *least precise* number in the problem.
- when multiplying or dividing two approximate numbers, the result should be rounded to the same number of significant digits as the number in the problem with the *fewest significant digits*.

Key Words and Phrases

significant digits	random errors	systematic errors
absolute error	relative error	accuracy
precision		

Key Concepts and Skills

- determine the number of significant digits in a given number.
- distinguish between random errors and systematic errors.
- compute absolute and relative errors.
- understand the difference between accuracy and precision.
- determine the precision of the sum or difference of two approximate numbers.
- determine the precision of the product or quotient of two approximate numbers.

Important Review Boxes

- A Brief Review of Rounding

Unit 3D Index Numbers: The CPI and Beyond

Overview

Suppose you want to compare the price of gasoline today with its price 25 years ago. Simply comparing the current price with the old price does not give a realistic comparison, because the price of just about everything increased during those 25 years. In this unit, we study the use of index numbers to make such comparisons more realistic. Among the most familiar index number is the **Consumer Price Index** (CPI), which essentially measures the cost of living and its changes due to **inflation**. There are many other index numbers, some of which will be explored in this unit, and all can be summarized by the following rule.

An **index number** provides a simple way to compare measurements made at different times or in different places. A value at one particular time (or place) is chosen as the *reference value*, and the *index number* for any other time (or place) is given by

$$\text{index number} = \frac{\text{value}}{\text{reference value}} \times 100.$$

Prices can be adjusted for inflation by applying the following principle: given a price in dollars for year X ($\$_x$), the equivalent price in dollars for year Y ($\$_Y$) is price in $\$_Y = (\text{price in } \$_X) \times \dfrac{\text{CPI}_Y}{\text{CPI}_X}$, where X and Y represent years, such as 1992 and 2007.

Key Words and Phrases

index number	reference value	Consumer Price Index
inflation	rate of inflation	price adjustment

Key Concepts and Skills

- compute index numbers and use them to make comparisons.
- use the CPI to compare prices.
- adjust prices for inflation.

Unit 3E How Numbers Deceive: Polygraphs, Mammograms, and More

Overview

The topics in this unit may seem a bit unrelated, but they are connected by the common theme of how numbers can be deceptive. The unit begins with a curious phenomenon known as **Simpson's paradox**. It can occur in many different ways, but it always arises when quantities are averaged and some crucial information is missing. The text examples illustrate the effect in many ways and should be studied carefully.

We next turn to surprisingly deceptive results that occur when using percentages in what are called *two-by-two tables* (such as Table 3.8). These tables are used to describe two different outcomes (for example, cured and not cured) for two different groups of people (for example, treatment and no treatment). Situations such as these give rise to the terms **false positive, false negative, true positive, and true negative**. Several examples are given as they apply to medical tests, lie detectors and drug tests. This is a very practical unit that contains some very subtle ideas.

Key Words and Phrases

Simpson's paradox	false positive	false negative
true positive	true negative	

Key Concepts and Skills

- identify and explain Simpson's paradox.
- compute percentages in tables and interpret them correctly.
- understand the subtleties in interpreting medical, drug, and polygraph tests.

4   Managing Your Money

Overview

One of the most immediate ways in which mathematics affects every person's daily routine is through finances: bank accounts, credit cards, loans, investments, and income taxes. In this chapter we take an in-depth look at these personal financial matters. In addition, we will explore the financial affairs of the United States as we look at the federal budget.

Unit 4A Taking Control of Your Finances

Overview

The key to your financial success is to approach all of your financial decisions with a clear understanding of the available choices. This unit focuses on maintaining a personal **budget**, for without a clear picture of your budget and monthly **cash flow**, you cannot effectively plan for the future.

A four-step budget is outlined in the text. Following this simple procedure allows you to take control of your personal finances. The four steps are:
- list your monthly income, including prorated amounts for income that is not received on a monthly basis.
- list *all* your monthly expenses, including prorated amounts for occasional expenses.
- compute your cash flow.
- make adjustments as needed to improve your cash flow.

Monthly *cash flow* is computed by subtracting your monthly expenses from your monthly income. With a positive cash flow, you'll have money left over at the end of the month (which can be applied toward savings); with a sustained negative cash flow, you are headed toward financial ruin.

Key Words and Phrases

> budget   cash flow

Key Concepts and Skills

- understand the importance of controlling your finances.
- know how to make a budget and compute cash flow.
- be aware of your bank balance, know what you spend, and don't overspend your budget.

Unit 4B The Power of Compounding

Overview

The phenomenon of **compounding** plays a fundamental role in all of finance, both on the investment side of the coin (earning and saving money) and on the loan side (borrowing money). The unit opens by considering simple and compound interest problems as they arise in banking. Compound interest problems rely on four pieces of information:

- **starting principal**, which we call P,
- **annual percentage interest rate**, which we call APR,
- **number of compoundings per year**, which we call n, and
- **number of years** the account is held, which we call Y.

With these variables defined, the compound interest formula is written as

$$A = P\left(1+\frac{\text{APR}}{n}\right)^{(nY)}.$$

This formula (and others in the chapter) must be evaluated on a calculator. Be sure you study the highlighted boxes that show you how to use your calculator effectively; needless to say, practice helps immensely.

An important distinction must be made between the APR and the **annual percentage yield** (APY) of a bank account. If an account compounds interest more than once per year, then the balance will increase by *more* than the APR in one year (due to the power of compounding). The percent by which the balance in the account actually increases in a year is the APY.

The more often compounding takes place during the year, the more the balance increases. The limiting case occurs when compounding takes place every instant, or **continuously**. With continuous compounding, you get the maximum return on your money (for a fixed APR). The continuous compounding formula, $A = P \times e^{(\text{APR} \times Y)}$, involves the mathematical constant e, which is approximately 2.71828. You should become familiar with how to compute various powers of e on your particular calculator.

The unit concludes with a very practical problem. The usual compound interest formulas tell you how much money you will have in your account after *Y* years with a given initial deposit *today*. But what if you know you would like to have, say \$30,000, in 20 years? How much should you deposit today in order to reach this goal? This is an example of a **present value** problem, and it's quite important for planning purposes.

Key Words and Phrases

simple interest	principal	compound interest
annual percentage rate (APR)	compound interest formula	annual percentage yield (APY)
continuous compounding	present value	

Key Concepts and Skills

- determine the balance in an account with simple interest given an initial deposit and interest rate.
- determine the balance in an account with compounding given an initial deposit, APR, number of compoundings, and number of years.
- determine the balance in an account with continuous compounding given an initial deposit, APR, and number of years.
- understand the difference between APR and APY, and be able to compute the APY for an account.
- determine the present value for an account that will produce a given balance after a specified number of years.

Important Review Boxes

- A Brief Review of Three Basic Rules of Algebra

Unit 4C Savings Plans and Investments

Overview

An **annuity** is a special kind of savings plan in which deposits are made regularly, perhaps every month or every year. People who are creating a college fund or building a retirement plan usually use an annuity. An annuity account increases in value due to compounding (as in regular bank accounts studied in the previous unit) and due to the regular deposits. Not surprisingly, the mathematics of annuity plans follows naturally from the compound interest formulas of the previous unit.

There is just one basic formula that needs to be mastered in order to work with annuities:

$$A = \text{PMT} \times \frac{\left[\left(1 + \dfrac{\text{APR}}{n}\right)^{(nY)} - 1\right]}{\left(\dfrac{\text{APR}}{n}\right)}.$$

This formula uses the variables needed for the compound interest formula of the previous unit (APR, number of compoundings, and number of years), *plus* the amount of the regular deposits (PMT). To keep matters simple (but still realistic) we assume that the regular deposits are made as often as interest is compounded. There is no doubt that the savings plan formula is complicated. Be sure to study the *Using Your Calculator* box to learn how to manipulate your calculator when evaluating the formula.

Just as with a bank account, it is practical to ask present value questions with savings plans. If you know how much money you would like to have at a future time (perhaps the time of retirement), then how much should you deposit monthly between now and then in order to reach that goal? Several practical examples of present value problems are presented in the text.

Having discussed the mathematics of investments, the unit concludes with a comprehensive summary of various types of investment plans. Some key considerations in choosing investment plans are **liquidity** (how easily your money can be withdrawn), **risk** (how safe your money is), and **return** (the amount that you earn on your investment). Liquidity is fairly easy to determine, risk is less easy to assess, and for return, there are simple formulas that give the total return and annual return on an investment:

$$\text{total return} = \frac{(A - P)}{P}, \quad \text{and} \quad \text{annual return} = \left(\frac{A}{P}\right)^{(1/Y)} - 1.$$

We examine four general types of investments in light of these three factors:
- small company stocks,
- large company stocks,
- corporate bonds, and
- Treasury bills.

These four types of investment have very different characteristics in terms of liquidity, risk, and return.

An essential part of investing money is keeping track of how well your investments are doing. Many people follow their investments using the financial pages of the newspapers. For this reason, we spend some time discussing how to read the financial pages for stocks, bonds, and mutual funds. Bonds in particular have some special terminology (such as **face value**, **coupon rate**, and **discount**) and a special formula for calculating the **current yield** on a bond.

Key Words and Phrases

savings plan	annuity	savings plan formula
stocks	bonds	cash
mutual fund	liquidity	risk
return	total return	annual return
portfolio	corporation	shares
market price	dividends	PE ratio
percent yield	face value	coupon rate
maturity date	discount	points
load	current yield	

Key Concepts and Skills

- determine the value of a savings plan given the regular deposit, APR, number of compoundings, and number of years.
- determine the present value for a savings plan that will yield a given balance after a specified number of years.
- understand the terms liquidity, risk, and return as they apply to various forms of investment.
- determine the annual yield on an investment given the total return.
- understand the terms used to report stocks in the financial pages of a newspaper.
- understand the terms used to report bonds in the financial pages of a newspaper.
- understand the terms used to report mutual funds in the financial pages of a newspaper.
- compute the current yield of a bond.

Important Review Boxes

- A Brief Review of Algebra with Powers and Roots

Unit 4D Loan Payments, Credit Cards, and Mortgages

Overview

While most people have bank accounts or investments that earn money, the reality is that most people also have debts due to loans of one kind or another. In this unit we will explore the most common kinds of loans: short-term loans (such as automobile loans), loans on credit cards, and **mortgages** (or house loans).

One basic formula governs all loan problems; it is called the **loan payment formula**:

$$PMT = \frac{P \times \left(\dfrac{APR}{n}\right)}{\left[1 - \left(1 + \dfrac{APR}{n}\right)^{(-nY)}\right]}.$$

This formula uses the now familiar variables of the previous units, though with different financial terms attached to some of them.

- PMT = regular payment amount,
- P = the initial amount of the loan (called the **loan principal**),
- APR = the interest rate on the loan,
- n = the number of payment periods per year, and
- Y = the loan term (the number of years over which it will be paid back).

Once again, the loan payment formula is rather complicated, so be sure to study the *Using Your Calculator* box.

The entire unit is devoted to using the loan payment formula for various practical problems. There are a lot of strategic questions involved in choosing a loan. Often you must choose between a loan with a short term and a loan with a longer term (it is typical to find that longer term loans have slightly higher interest rates). The best choice is not always clear, though the examples provide a good starting point in the decision making process.

For most people, a mortgage is the largest loan entered into over the course of a lifetime. Since mortgages involve large amounts of money over long periods of time, the decisions you make in choosing a mortgage are critical. The options and strategies involved with mortgages are explored (specifically, the differences and advantages of fixed-rate mortgages over adjustable rate mortgages). Also discussed are issues of refinancing, points, and closing costs. All in all, this is a practical unit, whose lessons will serve you well.

Key Words and Phrases

loan principal	installment loan	loan payment formula
mortgage	down payment	fixed rate mortgage
adjustable rate mortgage	closing costs	points
prepayment penalties	refinance	

Key Concepts and Skills

- know the terminology associated with loans.
- determine the loan payment given the loan principal, the APR, the number of payment periods, and the term of the loan.
- analyze two loan options to determine which is best for a given situation.
- be familiar with the practical issues associated with a home mortgage, such as points, closing costs, and fixed rate vs. adjustable rate loans.

Unit 4E Income Taxes

Overview

Another aspect of most peoples' financial lives is income taxes. The mathematics of income taxes is not difficult. However there are a lot of concepts and terminology involved with the subject. This unit is designed to give you a fairly thorough survey of income taxes, enough for you to make sense of complex tax forms and make critical decisions.

The first step in determining the amount of income tax you owe is to calculate your **taxable income**. From your **gross income** (the total of all your income), you subtract deductions and exemptions to arrive at your taxable income. Having computed your taxable income, the next step is to determine the tax itself. The U.S. tax system uses tax brackets or **marginal tax rates**: the percentage of your taxable income that you pay depends on your tax bracket. It also depends on your **filing status**: whether you are filing as a single person, a married person, or a head of household. This is also an appropriate place to discuss the **marriage penalty**.

Two other forms of income tax are **social security taxes** and **medicare taxes** (collectively called FICA taxes). These taxes apply to income from wages (not investments), and for self-employed people, these taxes must be paid in full by the taxpayer. The rules that govern these taxes are discussed with examples.

Of great political and personal interest are the taxes assessed on **capital gains**, profits made from selling property or stocks. We will examine the laws for capital gains and give several examples that illustrate the laws. Also of practical value are the tax implications of **tax-deferred savings plans** that many people use for retirement pensions and the **mortgage interest tax deduction,** which is a large tax benefit for people holding mortgages.

In this unit you will find some relief from the big formulas and long calculations of previous units in this chapter. This chapter is very conceptual and it presents a lot of new terminology and practical information.

Key Words and Phrases

gross income	adjusted gross income	deductions
exemptions	dependents	standard deduction
itemized deduction	taxable income	tax credits
progressive income tax	marginal tax rates	filing status
social security tax	medicare tax	capital gain
ordinary income	short-term capital gain	long-term capital gain
tax-deferred savings plan	mortgage interest deduction	marriage penalty

Key Concepts and Skills

- understand the basic terminology associated with income tax.
- compute taxable income from gross income given deductions and exemptions.
- use marginal tax rates to compute the tax on a given taxable income.
- compute the social security and Medicare taxes on a given income.
- understand and determine taxes on capital gains.
- incorporate tax-deferred savings plans and mortgage interest deductions into a tax calculation.

Unit 4F Understanding the Federal Budget

Overview

Several times in this book, particularly in problems about large numbers, we have referred to the federal budget or the federal debt. There is no question that the federal budget is a source of truly huge numbers – numbers so large that most people, including politicians, don't even have a sense of their size.

In this chapter, we give a brief survey of the essentials of the federal budget. While the details can be confusing, the general principles of the budget are straightforward. The government has **receipts** or **income** (primarily from taxes) and it is has **outlays** or **expenses** (that cover everything from defense to education to social security). In a given year, if outlays exceed receipts, then the government has a **deficit**. On the other hand, if receipts exceed expenses, then the government has a **surplus**. If the government has a deficit then it must borrow money and in the process goes into **debt**. The devastating effect of going into debt is that **interest** must be paid on the borrowed money. This interest payment itself becomes an expense category in the budget. The U.S. government typically spends around 10% of its budget on interest on the debt. If the government has many consecutive deficit years (as has been the case in the United States), then the debt continues to grow and cannot be paid off until there are surplus years.

In this unit we clarify the ideas of deficit and debt by looking at the budget of an imaginary small business. We then consider the federal budget for the last several years and see how the major spending categories have contributed to a deficit in each year. With many deficit years, it also becomes clear how the debt has grown to its current level of about $9 trillion.

The problems with the Social Security system are explored, and these raise a subtle point in federal budget reports. If you look closely at the federal budget summaries in this unit, you will see that the increase in the debt from one year to the next is *not* equal to the deficit for the current year, as one might expect. The explanation lies in the distinction between the **publicly held debt** and **gross debt**, and the way in which the U.S. government divides its **unified**

budget into two parts: the **off-budget** component (i.e. the Social Security system), and the **on-budget** component.

A balanced budget can be achieved either by increasing receipts or decreasing outlays or both. However, not all outlays can be reduced: **mandatory** expenses must be paid and **entitlements** are very difficult to cut. This leaves only **discretionary** expenses that are eligible for cuts.

The federal budget, together with its deficits, surpluses, and debts, affects everyone in many ways. This unit will give you a basic understanding of the important issues.

Key Words and Phrases

receipts	outlays	net income
deficit	surplus	debt
mandatory outlays	entitlements	discretionary outlays
publicly held debt	gross debt	unified budget
on budget	off budget	

Key Concepts and Skills

- explain the terms receipts, outlays, surplus, deficit, and debt.
- given a budget for a company or government, compute surplus or deficit.
- explain the difference between publicly held debt and gross debt.
- understand the major issues concerning the future of Social Security.

---

5   Statistical Reasoning

Overview

Much of the quantitative information that flows over us every day is in the form of data that are gathered through surveys and other statistical studies. When a person, business, or organization wants to know something, often the first idea that comes to mind is "collect data." From news reports to scientific research, from political polls to TV viewer surveys, we are surrounded by statistics. How are these numbers gathered? Are the conclusions based on those numbers reliable? And most important of all, should you believe the results of a statistical study? Should you change your life based on the results of a statistical study?

These are the questions that we will *begin* to answer in this chapter. The emphasis of this chapter is on qualitative issues, and there will not be a lot of computation involved. We will resume the study of statistical problems in Chapter 6 when we look at the quantitative side of statistics. If you master these two chapters, you will have a good foundation in statistical studies — enough to allow you to read the news critically and to take a complete course in statistics.

Unit 5A Fundamentals of Statistics

Overview

The goal of this unit is to learn how a statistical study should (and should *not*) be done. Of utmost importance is the distinction between the **population**, the group of people or objects that you would like to learn about, and the **sample**, the group of people and objects that you actually measure or survey. There are basically two aspects to any statistical study:
- collecting data from the people or objects in the sample, and
- drawing conclusions about the entire population based on the information gathered from the sample.

There are many ways that either of these steps can go wrong. Most of this chapter will deal with the first step, which is often called **sampling**. This step is crucial; if a sample is not chosen so that it is representative of the entire population, the conclusions of the study cannot be reliable.

Four different sampling methods are discussed:
- simple random sampling
- systematic sampling
- convenience sampling
- stratified sampling

It is also important to distinguish between two basic types of statistical studies. An **observational study** is one in which participants (people or objects) are questioned, observed, or measured, but in no way manipulated. By contrast, in an **experiment**, the participants *are* manipulated in some way, perhaps by dividing them into a **treatment group** and a **control group**. A subtle borderline situation is a **case-control study**, in which participants naturally form two or more groups by choice (for example, people who choose to smoke and those who choose to abstain). While a case-control study looks like an experiment, it is actually an observational study.

Other features of statistical studies that should be noted are the **placebo effect** (when people who *think* they received a treatment respond as if they received the treatment, even though they did not), **single blinding** (when participants do not know whether they receive a treatment), and **double blinding** (when neither researchers nor participants know who received a treatment).

You often hear statements like, "The President's approval rating is 54% with a margin of error of 5%." Because results of surveys are often given in this way, we will also look briefly at the notions of **confidence interval** and **margin of error**. Although these ideas are rather technical, it is important to understand how to interpret them. In Chapter 6, we will see how confidence intervals and margins or error are actually determined.

Key Words and Phrases

sample	population	raw data
sample statistics	population parameters	representative sample
simple random sampling	systematic sampling	convenience sampling
stratified sampling	bias	observational study
experiment	treatment group	control group
placebo	placebo effect	single blind experiment
double-blind experiment	case control study	cases
controls	margin of error	confidence interval

Key Concepts and Skills

- understand the distinction between a sample and the population.
- know the five basic steps of a statistical study.
- be able to form a representative sample by various sampling methods.
- understand the difference between an observational study, an experiment, and a case-control study.
- understand the placebo effect, and single and double blinding.
- interpret margins of error and confidence intervals.

Unit 5B Should You Believe a Statistical Study?

Overview

Suppose you read that 68% of all TV viewers watched the final game in the NCAA basketball tournament, or that 1 in 5 people in the world do not have access to fresh drinking water, or that a pre-election poll states that the Republican candidate leads by 5 percentage points. In this unit, we address a basic question: how do you know whether to believe the claims of these statistical studies?

The unit takes the form of a list of eight guidelines for evaluating a statistical study. Each guideline is accompanied by examples and analyses. It may not be possible to assess a particular study in light of *all* eight guidelines, but the more of these guidelines that a study satisfies, the more confident you can be about the conclusions of the study. The eight guidelines are listed below.

1. Identify the goal, type and population of the study.

2. Consider the source.

3. Look for bias in the sample.

4. Look for problems in defining or measuring the variables of interest.

5. Watch for confounding variables.

6. Consider the setting and wording in surveys.

7. Check that results are presented fairly.

8. Stand back and consider the conclusions.

Key Words and Phrases

bias   variables   selection bias

participation bias

Key Concepts and Skills

- understand the eight guidelines for evaluating a statistical study, and be able to carry them out on a particular study.

Unit 5C Statistical Table and Graphs

Overview

We often think of data and statistics as long lists of numbers, and indeed they start in that form. But these numbers don't become meaningful until they are summarized in some digestible form, and one of the best ways to represent long lists of numbers is with a picture. In this and the next unit, we will explore the many ways in which quantitative information can be displayed. This unit focuses on the basic types of graphs, ones that can often be drawn with a pencil and paper. The next unit highlights the more exotic types of graphs which are more difficult to produce, but equally common in news and research reports.

Often the best first step in summarizing data is to make a **frequency table** that shows how often each category in a study appears (for example, the number of students receiving grades of A, B, C, D, and F on an exam). Such a table can also show the **relative frequency** and the **cumulative frequency** for each category.

The type of display used for a set of data also depends on the type of data that are collected. We distinguish between **qualitative data** (non-numerical categories that have no natural order) and **quantitative data** (numerical values that are ordered).

Here is a summary of the types of graphs that will be considered in this unit.

- **Bar graphs** are used to show how some numerical quantity (for example, population) varies from one category to another (for example, for several different countries). The categories are usually qualitative and can be shown in any order.
- **Pie charts** are used to show the fraction (or percentage) of a population that falls into various categories. The categories are usually qualitative and can be shown in any order. For example, a pie chart could be used to show the fraction of a class that has brown, black, blond, or red hair.
- **Histograms** are bar graphs in which the categories are quantitative, and thus have a natural order. Histograms often show how many objects or people are in various categories. For example, a histogram would be used to show the number of people in a town that fall into age categories 0–9, 10–19, 20–29, and so on.
- **Line charts** serve the same purpose as histograms, but instead of using bars to indicate the number of objects or people in each category, they use dots connected by lines.
- **Time-series diagrams** are usually histograms or line charts that show how a quantity changes in time. For example, the day-to-day changes in the stock market could be displayed as a time-series diagram.

Key Words and Phrases

frequency table	relative frequency	cumulative frequency
bar graph	pie chart	histogram
line chart	time-series diagram	

Key Concepts and Skills

- construct a frequency table for a set of table showing frequencies, relative frequencies, and cumulative frequencies.
- determine an appropriate kind of display for a given set of data.
- display data with a bar graph by finding the correct height of the bars.
- display data with a pie chart by finding the correct angles for the sectors of the pie.
- display data with a histogram by finding the correct height of the bars.
- display data with a line chart by finding the correct location of points on the line chart.
- display data with a time-series diagram by finding the correct location of points on the diagram.

Unit 5D Graphics in the Media

Overview

Whereas the previous unit dealt with graphs that are relatively easy to produce yourself, this unit surveys more complicated displays that frequently appear in the media. The goal is not to actually produce these more sophisticated graphs; often a powerful software packages is needed to create them. Rather the emphasis will be on interpreting the information presented in a graphic.

Here is a summary of the types of graphs that we will consider in this unit.

- **Multiple bar graphs** are used to show how *two or more* quantities (for example, population and birth rate) vary from one category to another (for example, for several different countries). Multiple bar graphs are really two or more bar graphs combined with each other; they include multiple histograms.
- **Stack plots** are used to display several quantities simultaneously, often showing how they all change in time. For example, a stack plot could be used to show how the incidence of several diseases has changed in time.
- **Geographical plots**, such contour plots and weather maps, give a two-dimensional picture of how a quantity varies over a geographical region. Such displays are also used to show the distribution of a disease over a geographical region.
- **Three-dimensional** graphs take many forms, but they are all used to show how three different quantities are related to each other.

As graphical displays become more complex, there is more opportunity for confusion and deception. The unit closes by discussing several ways in which you can be misled by such displays — they must be interpreted with caution.

Key Words and Phrases

multiple bar graph	stack plot	three-dimensional graphics
contour plots	exponential scale	pictograph
perceptual distortions	percent change graphs	

Key Concepts and Skills

- interpret multiple bar graphs and histograms.
- interpret stack plots.
- interpret three-dimensional graphics.
- interpret contour plots.
- interpret percent change graphs.
- interpret exponential scales.
- detect deceptive pictographs or other displays of data.

Unit 5E Correlation and Causality

Overview

This unit deals with the critical task of determining whether one event *causes* another event. The process often begins by looking for **correlations** between two variables; a correlation exists when higher values of one variable are consistently associated with higher (or lower) values of the second variable. For example, body weight is correlated with height, because in general, greater body weights go hand-in-hand with greater heights. Correlations can be detected by making a **scatter diagram** of the two variables in question.

There are three possible explanations for a correlation between two variables. It may be due to

- a coincidence,
- a common underlying cause, or
- a genuine cause and effect relation.

Of the three possibilities, the most interesting and the most difficult to establish is the last one: when can we be sure that one event caused another?

The unit presents six methods (attributed to the philosopher John Stuart Mill) for establishing cause and effect relations:

1. Look for situations in which the effect is correlated with the suspected cause even while other factors vary.

2. Among groups that differ only in the presence or absence of the suspected cause, check that the effect is similarly present or absent.

3. Look for evidence that larger amounts of the suspected cause produce larger amounts of the effect.

4. If the effect might be produced by other potential causes (besides the suspected cause), make sure that the effect still remains after accounting for these other potential causes.

5. If possible, test the suspected cause with an experiment. If the experiment cannot be performed with humans for ethical reasons, consider doing the experiment with animals, cell cultures, or computer models.

6. Try to determine the physical mechanism by which the suspected cause produces the effect.

Because causality is important in legal cases, we also look at the legal standards for establishing causality. They are crucial in determining the guilt or innocence of defendants. They are

- possible cause,
- probable cause, and
- cause beyond reasonable doubt.

The unit also presents three case studies in which identifying a cause and effect relation led to a discovery or the solving of a mystery.

Key Words and Phrases

correlation	scatter diagram	positive correlation
negative correlation	coincidence	underlying cause
causality	probable cause	possible cause
cause beyond reasonable doubt		

Key Concepts and Skills

- construct a scatter diagram for two variables.
- identify positive, negative, and no correlation between two variables.
- understand the three explanations for a correlation.
- apply the six methods for determining causality.
- understand the three legal levels of confidence in causality.

---

6   Putting Statistics to Work

Overview

We now continue our study of statistics, having already introduced some of the qualitative aspects in Chapter 5. The treatment of statistics in this chapter is more quantitative and involves some numerical calculations. In Unit 6A we present several methods for characterizing data sets, including the calculation of the mean, median, and mode. The next unit continues in the same vein by discussing measures of variation in a data set. In Unit 6C, we introduce the normal distribution which lies at the heart of many statistical methods. The final unit discusses the powerful process known as *statistical inference* – drawing reliable conclusions about a population based on data taken from a sample of that population.

Unit 6A Characterizing a Data Distribution

Overview

Typical data sets often contain hundreds or thousands of numbers, so one goal is to summarize data sets in compact and meaningful ways. In this unit we explore methods used to describe and summarize data beyond the frequency tables discussed in the previous chapter.

For a very concise summary of a set of data, it is common to compute its **mean, median**, and/or **mode**. These one-number summaries are defined as follows:

- The *mean* of a data set is calculated by the formula

$$\text{mean} = \frac{\text{sum of all values}}{\text{total number of values}}.$$

- The *median* is the middle score in the data set. Note that there will be two "middle" values if a data set has an even number of data points; if the two middle values are different, the median lies halfway between them.

- The *mode* is the most common score in a data set. A data set may have more than one mode, or no mode.

Having determined where the "center" of the data set lies, it is next useful to describe the shape of the data: are the data values symmetric? Are they weighted to the right or left? These questions can be answered qualitatively by finding whether the data values are **positively** or **negatively skewed**.

Key Words and Phrases

distribution	mean	median
mode	single-peaked	bimodal
symmetric	outliers	positively skewed
negatively skewed		

Key Concepts and Skills

- compute the mean, median, mode of a set of data.
- understand the effect of outliers on mean and median.
- be aware of possible sources of confusion about averages.
- determine the qualitative shape of a data set and identify whether it is skewed.

Unit 6B Measures of Variation

Overview

We continue the theme of the previous unit by exploring additional ways to summarize a distribution of data in compact ways. In this unit, we look at methods to describe the *spread* or **variation** of a set of data. The simplest measure of variation is called the **range**, the difference between the maximum and minimum data values. We can improve on the range by computing the **upper** and **lower quartiles**, the mean, and the maximum and minimum data values to make a **five-number summary** for the distribution. A **boxplot** is a picture of the five-number summary and provides a good snapshot of the distribution.

The variation is best captured in a single number by computing the **standard deviation** of a distribution, shown here:

$$\text{standard deviation} = \sqrt{\frac{\text{sum of (deviations from the mean)}^2}{\text{total number of data values} - 1}}.$$

This calculation can become laborious for large data sets, and is often done with calculators or computer software. But you should know how to set up the standard deviation calculation using a table, as shown in Table 6.2. A useful approximation to the standard deviation is given by the **range rule of thumb**:

$$\text{standard deviation} \approx \frac{\text{range}}{4}.$$

Key Words and Phrases

range	five-number summary	upper quartile
lower quartile	box plot	deviation
standard deviation		

Key Concepts and Skills

- compute the range of a data distribution.
- find the lower and upper quartiles of a data distribution.
- construct the five-number summary and a boxplot for a data distribution.
- compute and interpret the standard deviation of a small data set.

Unit 6C The Normal Distribution

Overview

This unit is about the normal distribution, which plays an important role in all of statistics. This unit lays the groundwork for the following unit in which we learn how draw conclusions about an entire population when we have data from a relatively small sample of that population. This process of inferring something about the population based on a sample is often called **inferential statistics**.

We begin by introducing the **normal distribution**, a whole family of symmetric distributions that are characterized by their mean and their standard deviation. Many variables, such as test scores, heights, weights, and other physical characteristics, follow a normal distribution closely. One of the most important results about the normal distribution is the **68-95-99.7 Rule:**

- About 68% (closer to 68.3%) of the data points fall within 1 standard deviation of the mean.

- About 95% (closer to 95.4%) of the data points fall within 2 standard deviations of the mean.
- About 99.7% of the data points fall within 3 standard deviations of the mean.

Thus if we know that a variable is distributed normally and we know its mean and standard deviation, we can say a lot about where a particular data value lies in the distribution. Working with the normal distribution is made much easier with the use of **standard scores** (or *z-scores*) and **percentiles**. The standard score simply measures how many standard deviations a particular data value is above or below the mean. Once the standard score is known, the percentile can be found using Table 6.3 (which appears in all statistics books). Similarly, given a percentile, the corresponding *z*-score can be found.

Key Words and Phrases

normal distribution   68-95-99.7 rule   percentile

standard (*z*-) score

Key Concepts and Skills

- identify normal distributions and know how they arise.
- draw a normal distribution curve with a given mean and standard deviation.
- use the 68-95-99.7 Rule to analyze a normal distribution.
- convert standard scores to percentiles and vice versa.

Unit 6D Statistical Inference

This unit explains in a fairly qualitative way one of the most important aspects of statistics: the process by which information from a sample can be used to make reliable claims about the entire population from which the sample is taken. This is a technical topic, but it is so important that it's worth understanding, at least in general terms.

The unit begins with an explanation of **statistical significance**: a set of observations or measurements is said to be statistically significant if it is unlikely to have occurred by chance. For example, if we toss a fair coin 100 times and see 52 heads, we would not be surprised, and might attribute the outcome to chance. However, if we observed 80 heads in 100 tosses of a fair coin, it would be difficult to attribute this outcome to chance; we would say this outcome is statistically significant.

We can be more precise if we use probability in an intuitive way:
- if the probability of an observed outcome occurring by chance is 1 in 20 (0.05) or less, the outcome is significant at the **0.05 level of significance.**

- if the probability of an observed outcome occurring by chance is 1 in 100 (0.01) or less, the outcome is significant at the **0.01 level of significance**.

The discussion next returns to **margins of error** and **confidence intervals**. The problem of interest is very common for opinion polls and surveys. Suppose you want to know the proportion (or percentage) of people in a population who hold a particular opinion or have some trait. The usual practice is to select a sample and determine what proportion (or percentage) of people in the sample hold that particular opinion or have that trait; this is called the **sample statistic**. Knowing the sample statistic, what can you conclude about the entire population?

You will see in the unit (with plenty of explanation and examples) that if the sample has n individuals, then the proportion of interest for the population lies within $1/\sqrt{n}$ of the sample statistic — with 95% certainty. The quantity $1/\sqrt{n}$ is called the **margin of error** and the interval that extends $1/\sqrt{n}$ on either side of the sample statistic is called the **95% confidence interval**. For example, if in a randomly selected sample of $n = 10,000$ people, 45% have blond hair, then the margin of error is $1/\sqrt{10,000} = 0.01 = 1\%$ and the confidence interval is $45\% - 1\%$ to $45\% + 1\%$ or 44% to 46%. We can conclude with 95% certainty that between 44% and 46% of the entire population has blond hair. This is one of the most basic results of inferential statistics. It allows us to say something about an entire population based on what we learn from a sample.

The last topic in the unit is hypothesis testing, which is the procedure for determining whether a claim about a population, based on observations of a sample, is valid. The claims we will study involve either a population mean (for example, the mean weight of all college women is 125 pounds) or a population proportion (for example, 54% of all college students eat pizza at least once a week). The hypothesis test involves two hypotheses. The **null hypothesis** claims a specific value for a population parameter (it is often the value expected in the case of no special effect). It takes the form:

null hypothesis: population parameter = claimed value

The **alternative hypothesis** is the claim that is accepted if the null hypothesis is rejected.

There are two possible outcomes of a hypothesis test:
- rejecting the null hypothesis, in which case we have evidence that supports the alternative hypothesis.
- not rejecting the null hypothesis, in which case we lack sufficient evidence to support the alternative hypothesis.

How do we decide whether or not to reject the null hypothesis? Here's where the subject gets a bit technical to carry out in detail. But we can say it in words:
- If the chance of a sample result at least as extreme as the observed result is less than 1 in 100 (or 0.01), the test is significant at the 0.01 level.

The test offers strong evidence for rejecting the null hypothesis (and accepting the alternative hypothesis).

- If the chance of a sample result at least as extreme as the observed result is less than 1 in 20 (or 0.05), the test is significant at the 0.05 level. The test offers moderate evidence for rejecting the null hypothesis.
- If the chance of a sample result at least as extreme as the observed result is greater than 1 in 20, the test is not significant. The test does not provide sufficient grounds for rejecting the null hypothesis.

While this subject is challenging, the text does offer several good examples and analogies to clarify this important topic.

Key Words and Phrases

statistical significance	sampling distribution	confidence interval
margin of error	null hypothesis	alternative hypothesis
significance at the 0.05 level	significance at the 0.01 level	

Key Concepts and Skills

- explain the meaning of statistical significance.
- construct and interpret a confidence interval given a margin or error.
- describe the basic purpose and aspects of a hypothesis test.
- construct a null and alternative hypothesis for a given situation.
- given the relevant probability, decide whether or not to reject a null hypothesis.

7   Probability: Living With the Odds

Overview

Probability is involved in nearly every decision we make. Often it is used on a subjective or intuitive level; but occasionally we try to be more precise. As you will see in this chapter, probability is one of the most applicable branches of mathematics. Before we are done with the chapter, we will be able to apply probability to lotteries, gambling, life insurance problems, and air traffic safety. It is a fascinating chapter, full of mathematics and real-life problems.

Unit 7A Fundamentals of Probability

Overview

We all have an intuitive sense of what a probability is and that sense is useful. A probability is a number between zero and one. If the probability of an event is zero, then it cannot happen; if the probability of an event is one, then it is certain to happen; and if the probability is somewhere in between zero and one, it gives a measure of how likely the event is to happen.

We can approach probabilities in three different ways:

- a **theoretical probability** uses mathematical methods to find the probability of an event occurring.
- an **empirical probability** is determined with experiments or by data collection.
- a **subjective probability** is based on intuition.

This unit focuses on theoretical probability calculations, with a lighter treatment of empirical and subjective probabilities.

If you want to find the probability that a fair die will show a 5 when rolled once, you might reason that there are six possible **outcomes** of a single roll, and a success (rolling a 5) is one of those six outcomes. You (correctly) conclude that the probability of rolling a 5 is 1/6. This is the basic procedure for computing theoretical probabilities:

Step 1: Count the number of possible outcomes of an **event**.
Step 2: Count the number of outcomes that represent **success** — that is, the number of outcomes that represent the sought after result.
Step 3: Determine the probability of success by dividing the number of successes by the total number of possible outcomes:

$$\text{probability of success} = \frac{\text{number of outcomes that represent success}}{\text{total number of possible outcomes}}$$

This procedure can be used to compute probabilities in many different situations, as you will see in the examples of the text.

An often-used important fact is that if the probability of an event A occurring is $P(A)$, then the probability that it does *not* occur is $1 - P(A)$. For example, the probability of rolling a fair die and getting a five is 1/6; therefore the probability of rolling *anything but* a five is $1 - 1/6 = 5/6$.

Suppose you flip three coins and are interested in the number of heads that appear. In this case, there are four possible events: three heads, two heads, one head, and three tails. We will show how to find the probability of all of the events. The collection of the probabilities of *all* possible events is called a **probability distribution**.

There are many situations in which it's impossible to determine theoretical probabilities. In these situations, we can't use mathematical methods, but we have records or data to work with. Such an approach leads to empirical probabilities. For example, we often talk about a 100-year flood, which means that based on historical records, a flood of this magnitude occurs once every 100 years. So the empirical probability of such a flood occurring in any given year is 1/100. We discuss empirical probabilities because they are used so often in practice.

Finally, we take a few pages to clarify the confusion that can arise over the use of the word *odds*. The odds *for* an event A are found by dividing the probability of an event occurring by the probability of the event *not* occurring. For example, the odds of rolling a fair die and getting a five are *1 to 5*, because $(1/6) \div (5/6) = 1/5$. The odds *against* an event A are found by dividing the probability of an event not occurring by the probability that it does occur, which is just the reciprocal of the odds for an event. Thus the odds against rolling a five are *5 to 1*.

Key Words and Phrases

outcome	event	theoretical probability
empirical probability	subjective probability	probability distribution
odds		

Key Concepts and Skills

- distinguish between theoretical probability, empirical probability, and subjective probability.
- use the three-step process to determine simple theoretical probabilities.
- find probability distributions for coin and dice experiments.
- determine empirical probabilities from data or historical records.
- find the probability of an event *not* occurring.
- find the odds for an event from its probability.

Important Review Boxes

- A Brief Review of the Multiplication Principle

Unit 7B Combining Probabilities

Overview

In the previous unit, we studied methods for finding the probability of individual events. However, many interesting situations actually consist of multiple events. For example, what is the probability of tossing ten heads in a row with a fair coin? Or what is the probability of drawing a jack or a heart from a standard deck of cards? We will answer these and many other practical questions in this unit.

The unit presents five different techniques that involve multiple events. As always, it's important to know *how* to apply these techniques and *when* to apply them. The five situations covered by these methods are:
- independent AND events
- dependent AND events
- either/or mutually exclusive events (non-overlapping)
- either/or jointly possible events (overlapping)
- *at least once* rule for independent events.

The term **AND event** refers to two or more events *all* happening. For example, what is the probability of rolling two fair dice and seeing a six on *both* dice (a six *and* a six)? Or what is the probability of drawing four cards from a standard deck and getting an ace each time? The important distinction with AND events is whether they are **independent** or **dependent**. If the occurrence of one event does not affect the others (for example, in rolling two dice, the outcome of one die does not affect the outcome of the other die), then we have independent events. The rule that applies for two independent events is

$$P(A \; and \; B) = P(A) \times P(B).$$

This principle can be extended to any number of independent events. For example, the AND probability of A *and* B *and* C is

$$P(A \; and \; B \; and \; C) = P(A) \times P(B) \times P(C).$$

If the events are not independent, then a bit more care is needed. Suppose you draw two cards from a deck, but don't replace the first card before drawing the second. The outcome of the second card depends on the outcome of the first card. The rule that applies in this case for two dependent events is

$$P(A \; and \; B) = P(A) \times P(B \; given \; A).$$

With **either/or events**, we are interested in the probability of *either* event A *or* event B occurring. The important distinction with either/or events is whether the events in question are mutually exclusive. If the occurrence of the first event prevents the occurrence of the second event, then the events are **mutually exclusive** or **non-overlapping**. For example, when selecting one card from a standard deck, drawing a heart and drawing a diamond are mutually exclusive events because if one event occurs (say, drawing a heart), then the other event (drawing a diamond) cannot occur. The rule that applies in this case with two mutually exclusive events is

$$P(A \; or \; B) = P(A) + P(B).$$

This principle can be extended to any number of mutually exclusive events. For example, the probability that either A *or* B *or* C occurs is

$$P(A \; or \; B \; or \; C) = P(A) + P(B) + P(C).$$

If events are **overlapping** (for example, going into a room of people and meeting a woman *or* a Democrat), then a modification of the above rule must be used. The rule that applies in this case is

$$P(A \ or \ B) = P(A) + P(B) - P(A \ and \ B).$$

In general, for AND probabilities we multiply probabilities and for either/or probabilities we add probabilities. However, modifications must be made in the cases of dependent or overlapping events.

Finally, a very important situation arises when we ask about the probability of an event happening *at least once* in a string of several independent trials of an experiment. For example, if you buy ten lottery tickets, what is the probability of *at least* one ticket being a winner? If you roll a die ten times, what is the probability of rolling *at least* one six? The rule that gives the probability of an event A occurring *at least once* in n trials is

$$P(A \text{ at least once in } n \text{ trials}) = 1 - P(A \text{ does not occur in } n \text{ trials})$$

$$= 1 - P(not \ A)^n.$$

The unit has explanations of these rules and many examples to show how they are used.

Key Words and Phrases

AND probability	independent events	dependent events
either/or probability	overlapping events	non-overlapping events
at least once rule		

Key Concepts and Skills

- identify AND probabilities and either/or probabilities.
- distinguish between independent events and dependent events.
- distinguish between non-overlapping events and overlapping events.
- identify *at least once* situations.
- know when and how to apply the five probability rules given in the unit.

Unit 7C The Law of Large Numbers

Overview

This unit is designed to strengthen your intuition about probabilities and to point out some common misconceptions about probabilities. It also deals with the concept of expected value, which has many important applications.

We start by discussing the **law of large numbers**. We know that the probability of tossing a head with a fair coin is 1/2. This does not mean that when you toss a coin twice, you will always get one head and one tail. Nor does it mean that when you toss a coin ten times, you will always get five heads and five tails. What we *can* say is that the more often you toss the coin, the closer the proportion of heads is to the probability of finding heads on a single coin toss. This is an example of the law of averages.

The important concept in this chapter is **expected value**. Consider a situation in which there are several different outcomes. Each outcome has a known probability and a known payoff, or **value** (which could be positive or negative). For example, a lottery may have a $10 prize, a $100 prize, a $1000 prize, and a grand prize of $1,000,000. These are the values (payoffs) of the four outcomes, and each outcome has a certain probability. If you play this lottery many times, how much do you expect to win or lose in the long run? The answer is given by the expected value.

There is a general rule for computing expected values. In a situation with just two outcomes, the expected value is given by

$$\text{expected value} = \begin{pmatrix} \text{value of} \\ \text{event 1} \end{pmatrix} \times \begin{pmatrix} \text{probability of} \\ \text{event 1} \end{pmatrix} + \begin{pmatrix} \text{value of} \\ \text{event 2} \end{pmatrix} \times \begin{pmatrix} \text{probability of} \\ \text{event 2} \end{pmatrix}.$$

The unit presents several other applications of expected value as it arises in life insurance policies, lotteries, and the **house edge** in casino gambling. There are some important practical lessons to be learned here.

The law of averages is related to the principle called the **gambler's fallacy**. Gamblers often feel that if they get behind or start losing money, then their luck will change and their chances of winning will improve. The fact is that probabilities do not change just because someone is losing. Once behind, it is likely that you will stay behind or get even further behind.

Key Words and Phrases

law of large numbers   expected value   gambler's fallacy

house edge

Key Concepts and Skills

- explain the law of averages.
- explain the gambler's fallacy.
- compute the expected value in situations with two or more outcomes.
- understand the house edge and compute it in specific situations.

Unit 7D Assessing Risk

Overview

This unit is a nice collection of topics, all related in some way to probability. The first topic is estimating risk. You often hear statements such as "you are more likely to get struck by lightning than die in an airplane crash." How are such statements concocted? And to what extent are they true? We will explore how such comparisons of risk are made, particularly as they apply to **vital statistics** — data about births and deaths.

You will see that these problems are examples of empirical probability. If we know the number of deaths due to, say, heart disease, we can estimate the probability of a person dying from heart disease. If we know the number of deaths due to automobile accidents, we can estimate the chances of a person dying in an automobile accident. With these two facts, it is possible to compare the risks of automobiles and heart disease. Clearly, risk analysis and decision-making are closely related.

Related to vital statistics are the subjects or **mortality** and **life expectancy**. We will look at graphs of death rates and life expectancy as they change with age. The interpretation of these graphs reveals a few surprises.

Key Words and Phrases

accident rate	death rate	vital statistics
life expectancy	mortality	

Key Concepts and Skills

- determine the probability of an event given the number of people who experience the event.
- understand different measures of risk.
- compare the risks of two events or two causes of death.
- interpret life expectancy and mortality tables.

Unit 7E Counting and Probability

Overview

We all know how to count *individual* objects or people, but there are many other types of counting problems that arise when we want to count *groups* of objects or people. That is the subject of this unit. There are four basic counting methods presented in this chapter:

- selections from two or more groups (multiplication principle),
- arrangements with repetition,
- permutations (where repetition is not allowed), and
- combinations (again, repetition is not allowed).

In this unit, we will learn not only how to apply these methods, but equally important, how to determine *when* to use each method.

Selection from two or more groups occurs when we have several different groups of objects and we need select one item from each group. For example, in a restaurant, you might have four choices for an appetizer (first group), six choices for a main course (second group), and five choices for a dessert (third choice). How many different meals could you select?

Problems of this sort can be solved using a table, a tree, or (the easiest of all) the **multiplication principle** (see Review Box in Unit 7A). The total number arrangements of items from two or more groups is

> (number of items in first group) × (number of items in second group) ×
> (number of items in third group) × ···· × (number of items in last group).

In the above example, there would $4 \times 6 \times 5 = 120$ different possible meals that you could order.

Arrangements with repetition occur when you select items from a single group and items may be used more than once. Often the items in problems of this kind are letters or numerals. For example, how many different three-digit area codes can be formed from the numerals 0 through 9? There are 10 ways to choose the first numeral and, because repetition is allowed, there are 10 ways to choose the second numeral, and 10 ways to choose the third numeral. This amounts to a total of $10 \times 10 \times 10 = 1000$ different three-digit area codes. The general rule is that

> if we make r selections (for example, the three digits of the area code) from a group of n items (for example, the numerals 0–9), n^r different arrangements are possible.

Permutations also involve selecting items from a single group with one important difference — repetition is not allowed. Furthermore, in counting permutations, the order of the arrangements matters; that is, we count ABCD and DCBA as two different arrangements (or permutations). To summarize, permutations require

- selection from a single group,
- repetition is not allowed, and
- order matters.

In the unit, we work through several examples towards the general formula for counting permutations. We need a bit of mathematical notation along the way, so we introduce the **factorial** function.

$$n! = n \times (n - 1) \times (n - 2) \times \ldots \times 2 \times 1.$$

With factorials in hand, we can write a general formula for counting permutations.

> If we make r selections from a group of n items, the number of possible permutations is
> $$_nP_r = \frac{n!}{(n - r)!},$$
> where $_nP_r$ is read as "the number of permutations of n items taken r at a time."

Combinations are much like permutations with one difference — order of selection does not matter; that is, we count ABCD and DCBA as the *same* combination. The requirements for combinations are

- selection from a single group,
- repetition is not allowed, and
- order does *not* matter.

The number of combinations of *n* objects taken *r* at a time is the same as the number of permutations, except that we have to correct for the overcounting embedded in the permutation formula (permutations involve a specific order and combinations don't). This subtle point is explained in the text. The result is the combinations formula.

If we make *r* selections from a group of *n* items, the number of possible *combinations*, in which order does not matter, is

$$_nC_r = \frac{_nP_r}{r!} = \frac{n!}{(n-r)! \times r!}$$

where $_nC_r$ is read as "the number of combinations of *n* items taken *r* at a time."

The key to mastering these methods is knowing not only *how* to use a particular method, but *when* to use it. There are plenty of practice exercises at the end of the unit; you should work as many as possible. Another helpful hint is to be certain you are using your calculator effectively. Most calculators compute factorials directly, and some have keys for permutations and combinations — it is in your best interest to become familiar with them.

The unit concludes with a brief look at the subject of **coincidence**. Should you be surprised during an evening of playing cards when you are dealt a 13-card hand with ten hearts? Many so-called coincidences are bound to happen to someone, but we are often surprised when they happen to us. The famous birthday problem at the end of the unit illustrates this principal. We also explore the phenomenon of **streaks**, particularly in sports, when a player repeats a certain success for many consecutive games.

Key Words and Phrases

selection from two or more groups	multiplication principle	arrangements with repetition
permutation	combination	coincidence

Key Concepts and Skills

- describe the four different counting methods discussed in the unit.
- determine which of the four methods applies in a counting problem.
- apply each of the four methods on appropriate problems.
- understand why not all coincidences should be surprising.

8   Exponential Astonishment

Overview

If you had to learn just one lesson from a quantitative reasoning course, it might well be the difference between exponential growth (the present chapter), and linear growth (studied in Chapter 9). Exponential growth and decay impact our everyday lives in many ways, but most people are not aware of its presence or its power. In this chapter, you will learn how bank accounts, populations, and radioactive waste are governed by exponential models.

Unit 8A Growth: Linear versus Exponential

Overview

The difference between linear and exponential growth is stated at the outset.

- *Linear growth* occurs when a quantity grows by the same *absolute* amount in each unit of time.
- *Exponential growth* occurs when a quantity grows by the same *relative* amount — that is, by the same *percentage* — in each unit of time.

Notice how these facts are related to the ideas of absolute and relative change studied in Unit 3A. This unit introduces the ideas surrounding exponential growth in a very intuitive fashion. We use three parables (*From Hero to Headless*, *The Magic Penny*, and *Bacteria in a Bottle*) that illustrate quite dramatically the power of exponential growth. The goal is to develop some intuition about exponential growth and understand doubling processes.

These lessons are summarized in the highlight box at the end of the unit:
- Exponential growth is characterized by repeated doublings. With each doubling the amount of increase is approximately equal to the *sum* of all preceding doublings.
- Exponential growth cannot continue indefinitely. After only a relatively small number of doublings, exponentially growing quantities reach impossible proportions.

Key Words and Phrases

linear growth   exponential growth   doublings

Key Concepts and Skills

- explain the difference between linear and exponential growth.
- identify whether a given growth pattern is linear or exponential.
- understand the implications of a doubling process.

Unit 8B Doubling Time and Half-Life

Overview

Exponential growth is characterized by a constant **doubling time**. If a quantity (for example, a population or a bank account) is growing exponentially, then it doubles its size during a fixed period of time, and it continues to double its size over that same period of time for as long as current growth rates continue. For example, if a tumor, growing exponentially, has a doubling time of two months, then it doubles its number of cells during the first two months, and doubles that number again during the next two months, and continues to double *every* two months. Knowing the doubling time essentially defines the growth pattern for all times.

We denote the doubling time with T_{double}. If we know T_{double} for a particular quantity that grows exponentially, then over a period of time t, the quantity will increase by a factor of

$$2^{t/T_{double}}.$$

If we know the doubling time and the initial value of a particular quantity that grows exponentially, then we can find its value at all later times. The new values at later times are given by

$$\text{new value} = \text{initial value} \times 2^{t/T_{double}}.$$

If we know that a quantity grows with a constant percentage growth rate (for example, 5% per year), then we know it grows exponentially and that it has a constant doubling time. This suggests that there should be a connection between the percentage growth rate P, and the doubling time. We use a specific example of an exponentially growing population to present a widely used formula that relates percentage growth rate to doubling time. It is called the **Approximate Doubling Time Formula** or the **Rule of 70**. It says that

$$T_{double} \approx \frac{70}{P}.$$

This formula is an approximation and works best when the percentage growth rate is small (say, less that 10%). For example, if a bank account grows at 4% per year, its balance will double in approximately 70/4 = 17.5 years.

Everything we learned about exponential growth has a parallel with exponential decay. For example, if a quantity decays exponentially at a rate of 5% per month, it *decreases* by 5% every month. A quantity that decays exponentially has a constant **half-life** — the period of time over which it decreases its size by 50% or one-half.

If we know T_{half} for a particular quantity that decays exponentially, then over a period of time t, the quantity will decrease by a factor of

$$\left(\frac{1}{2}\right)^{1/T_{half}}$$

If we know the half-life and the initial value of a particular quantity that decays exponentially, then we can find its value at all later times. The new values at later times are given by

$$\text{new value} = \text{inital value} \times \left(\frac{1}{2}\right)^{1/T_{half}}.$$

The **Approximate Half-Life Formula** also applies to exponentially decaying quantities. If a quantity decreases by $P\%$ per unit time, its half-life is given by

$$T_{half} = \frac{70}{P}.$$

As with the Approximate Doubling Time Formula, this half-life formula is approximate and works best when P is small (say, less than 10%).

For those who are curious, the unit closes with the *exact* doubling time and half-life formulas. These formulas are a bit more complicated and require the use of logarithms. They are exact for all percentage growth and decay rates, not just small ones.

Key Words and Phrases

doubling time	percentage growth rate	approximate doubling time formula
Rule of 70	half-life	percentage decay rate
approximate half-life formula	exact doubling time formula	exact half-life formula

Key Concepts and Skills

- identify the percentage growth rate from the description of an exponential growth process.
- find the doubling time from percentage growth rate.
- find the percentage growth rate from doubling time.
- determine the value of an exponentially growing quantity given the doubling time and initial value.
- identify the percentage decay rate from the description of an exponential decay process.
- find the half-life from percentage decay rate.
- find the percentage decay rate from half-life.
- determine the value of an exponentially decaying quantity given the half-life and initial value.

Important Review Boxes

- A Brief Review of Logarithms

Unit 8C Real Population Growth

Overview

In the previous units of this chapter, we have seen that exponential growth models allow only for rapid and continual growth. While this sort of growth is realistic for some populations (for example, bacteria, tumor cells, or small populations of animals), it cannot continue forever. Eventually lack of space or resources must limit growth.

In this short unit we look briefly at more realistic approaches to population modeling. We begin by discussing how the overall growth rate really consists of **birth rates** and **death rates** (and immigration rates). We then introduce the important concept of **carrying capacity** — the maximum number of individuals that the environment can sustain. Any realistic population model must account for the carrying capacity.

The next observation is that as a population grows, its percentage growth rate cannot remain constant, as it does in an exponential growth model. The **logistic growth model** uses a growth rate that actually decreases as the population grows, and reaches zero when the population reaches the carrying capacity. An alternative scenario, called **overshoot and collapse**, is also presented as a realistic population model. The unit closes with a discussion of estimating the carrying capacity of the Earth.

Key Words and Phrases

birth rate	death rate	carrying capacity
logistic growth model	overshoot	collapse

Key Concepts and Skills

- determine the exponential growth rate from birth and death rates.
- understand the limitations of exponential growth models.
- understand the assumptions and effects of a logistic growth model.
- determine growth rates for a logistic model.
- describe overshoot and collapse models.

Unit 8D Logarithmic Scales: Earthquakes, Sounds, and Acids

Overview

Closely related to exponential growth and decay models are **logarithmic scales** — these are just the other side of the exponential coin, in that they scale quantities based on powers (exponents).

Each of the three logarithmic scales explored in the text (the *earthquake magnitude scale*, or Richter scale; the *decibel scale* for sound; and the *pH scale* for measuring acidity) takes two equivalent forms, one expressed in terms of logarithms, and one expressed in terms of powers. Following are the defining equations for these scales.

The earthquake magnitude scale:

$$\log_{10} E = 4.4 + 1.5M \quad \text{or} \quad E = (2.5 \times 10^4) \times 10^{1.5M}.$$

The decibel scale for sound:

$$dB = 10 \log_{10}\left(\frac{\text{intensity of the sound}}{\text{intensity of the softest audible sound}}\right) \text{ or } \frac{\text{intensity of the sound}}{\text{intensity of the softest audible sound}} = 10^{(\text{loudness in dB}/10)}.$$

The pH scale:

$$pH = -\log_{10}[H^+] \quad \text{or} \quad [H^+] = 10^{-pH}.$$

One key idea to understand about logarithmic scales is that changing a scaled number by a multiplicative factor does *not* correspond to a change in the quantity the scale measures by the same multiplicative factor. For example, if the magnitude of an earthquake is doubled from 4 to 8, the energy E released is not doubled, but rather it goes up by a *much* larger factor (in this case, the energy released is about 1,000,000 times as great). The factor by which the measured quantity increases with corresponding increases in the scaled number depends upon how the logarithmic scale is defined.

An interesting aspect of the intensity of sound (measured by the decibel scale) is that it follows an **inverse square law**. This is just one of several quantities in nature that obey the inverse square law (others include gravity, and the brightness of light).

Key Words and Phrases

logarithmic scale	earthquake magnitude scale	decibel scale
pH scale	acid	base
neutral	acid rain	inverse square law

Key Concepts and Skills

- determine the energy released by an earthquake given its magnitude, and compare two earthquakes of different magnitudes.
- compare the intensity of sounds that have different decibel levels.
- understand how the inverse square law affects the intensity of sound at a distance.
- understand the relationship between pH and [H$^+$].
- explore the relationship between the pH scale and acid rain.

9   Modeling Our World

Overview

Just as a map or a globe is a model of the earth and a set of floor plans is a model of a building, we can also create mathematical models that represent real-world problems and situations. In fact, it might be argued that one of the fundamental goals of mathematics is to create mathematical models.

In this chapter, we introduce perhaps the most basic tool of mathematical modeling: the function. In fact, we have already encountered modeling and functions prior to this chapter. The financial formulas of Chapter 4 are examples are functions, and the exponential formulas of Chapter 8 are examples of mathematical models. The presentation of functions in this chapter may be different than those you have seen elsewhere. We start at the beginning and develop the idea of a function in a practical and visual way. Hopefully, in this way, whether you are seeing functions for the first time or not, you will become comfortable with this essential mathematical concept.

Unit 9A Functions: Building Blocks of Mathematical Models

Overview

This unit is brief and has a single purpose. Using words, a few definitions, pictures, and tables, the goal is to give a very qualitative idea of a **function**.

A function is a relationship between two quantities or **variables**. A function could describe how your height increases in time or how the temperature decreases with altitude. We call the variables in question the **independent variable** and the **dependent variable**. The terminology arises because we usually think of the dependent variable *changing with respect to* the independent variable; that is, if we make a change in the independent variable, it produces a change in the dependent variable.

There are three different ways to visualize or represent a function:
- with a data table of values of the two variables,
- with a graph (picture), and
- with an equation or formula.

In the remainder of this unit, we explore the first two ways of representing functions. The first approach (a data table) is straightforward and probably familiar. So we focus on graphing functions. The use of equations will be studied in the next unit.

Two concepts are important in graphing functions.

- The **domain** of a function is the set of values that both make sense and are of interest for the *independent variable*.

- The **range** of a function consists of the values of the *dependent variable* that correspond to the values in the domain.

If you can identify the domain and range of a function, then you have saved yourself a lot of work. You need only to graph those values of the independent variable in the domain and those values of the dependent variable in the range. Having defined the domain and range, we present several examples of functions and their graphs.

Key Words and Phrases

mathematical model	function	variable
independent variable	dependent variable	graph
coordinate plane	axis	origin
quadrants	domain	range

Key Concepts and Skills

- find the domain and range of a function given in table form.
- use the domain and range of a function to scale the axes of its graph.
- know when it makes sense to "fill in" between the points of a graph.
- use the four-step process to create a function's graph given a data table.

Important Review Boxes

- A Brief Review of The Coordinate Plane

Unit 9B Linear Modeling

Overview

From the previous unit, you probably appreciate that graphs of functions can take many different forms. In this unit we study one very special, but widely used, family of functions — those functions whose graphs are straight lines. Not surprisingly, these functions are called **linear functions**.

The most important property of a linear function is its **rate of change**. If the graph of two variables is a straight line, it means that a fixed change in one variable always produces the *same* change in the other variable. The rate at which the dependent variable changes with respect to the independent variable is called the rate of change of the function. For linear functions, the rate of change is constant and it is equal to the **slope** of the graph. This fundamental property for linear functions is summarized here, in two equivalent ways.

$$\text{slope of a linear graph} = \frac{\text{vertical } rise}{\text{horizontal } run}$$

$$\text{rate of change} = \text{slope} = \frac{\text{change in } dependent \text{ } variable \text{ from } P_1 \text{ to } P_2}{\text{change in } independent \text{ } variable \text{ from } P_1 \text{ to } P_2}.$$

Having established the equivalence of the slope and the rate of change, we introduce a rule for calculating the change in the dependent variable. This rule simply says that if we know the rate of change of a linear function and we are given a change in the independent variable, then we can determine the corresponding change in the dependent variable:

Change in dependent variable = (rate of change) × (change in independent variable).

The rate of change rule is really just a stepping-stone to the final goal of the unit — to write a general equation for a linear function. After a detailed example to motivate the idea of a linear equation, we present the general form of a linear equation. It looks like this:

dependent variable = initial value + (rate of change × *independent variable*)

Up to this point we use words for the variable names. This practice can become cumbersome. So for the sake of economy, we can use single letter names for variables. For example, instead of writing *time*, we just use *t*; and instead of writing *number of chips*, we could use *N*. Don't let the use of letters confuse you; it just makes working with linear equations easier.

After several examples of creating and using linear equations, we come to one last topic. A linear equation is a model or a compact description of a particular situation. Once we have it, it can be used for prediction or to answer other useful questions. We want to be able to answer questions such as:

- when the independent variable has a certain value, what is the corresponding value of the dependent variable?
- when the dependent variable has a certain value, what is the corresponding value of the independent variable?

This brings us to the necessity of *solving* linear equations when we are given a particular value of either the dependent variable or the independent variable. Two simple rules are all we need to solve any linear equation for either variable:

- we can always add or subtract the same quantity from both sides of an equation.
- We can always multiply or divide both sides of an equation by a (nonzero) quantity.

Several examples conclude this action-packed unit. There is a lot of material in this unit and it is best to work at it slowly and allow plenty of time for reading and for practice.

Key Words and Phrases

linear function	linear graph	rate of change
slope	initial value (*y*-intercept)	$y = mx + b$

Key Concepts and Skills

- determine the slope of a straight line given two points on the line.
- determine the rate of change of a linear function, either from a description of the function or from two points associated with the function.
- compute the change in the dependent variable given the rate of change and a change in the independent variable.
- find the equation of a linear function given the rate of change and the initial value.
- evaluate a linear function for the dependent variable given a value of the independent variable.
- solve a linear function for the independent variable given a value of the dependent variable.
- create a linear function from information about its rate of change and initial value.

Unit 9C Exponential Modeling

Overview

Having learned about the fundamentals of exponential growth and decay in Chapter 8, we can put these ideas to work to create models. Just as we used linear functions to model real world situations, we now use exponential functions to model situations in which exponential growth or decay occur. We begin by introducing the general **exponential function**

$$Q = Q_0 \times (1 + r)^t,$$

where t represents time. This equation requires an initial value Q_0 and a growth or decay rate r. Notice that if r is positive, then Q grows in time and if r is negative, then Q decays in time. Also important is that the units of time used for r and t must be the same (for example, if r is 5% *per day*, then t is measured in days). Once a specific exponential function is found, it can be used to predict the value of Q at all future times.

Be sure to study the highlight box entitled *Forms of the Exponential Function*. It shows that there are really several forms for the exponential function depending on whether you are given a growth rate, a doubling time, or a half-life.

With these exponential functions at hand, the rest of the unit is devoted to various applications. We look at how population growth, economic inflation, oil consumption, pollution, and drugs in the blood can all be modeled using these laws. Of particular importance is the technique called **radiometric dating**, which also relies on the exponential decay law. If these examples don't convince you of the widespread presence of exponential growth and decay, you will find even more applications in the problems.

Key Words and Phrases

exponential function   growth or decay rate   radiometric dating

Key Concepts and Skills

- given either a growth rate or a doubling time, use the appropriate form of the exponential function to model an exponentially growing quantity.
- given either a decay rate or a half-life, use the appropriate form of the exponential function to model an exponentially decaying quantity.
- be familiar with forms of the exponential function that use the doubling time or half-life.
- understand radiometric dating and know how to determine the age of a material that contains a radioactive element.

Important Review Boxes

- A Brief Review of Algebra with Logarithms

10   Modeling with Geometry

If you like geometry, you will probably like this chapter. If you didn't, this chapter may show you some things you had not thought about before and may be able to use. The first unit is a review of some fundamental concepts from high school geometry.

The last two sections have applications many of you have not seen before. While the ideas in Unit 10B are not difficult, they are particularly useful in many walks of life. Applications to astronomy, road construction, and how tall to make a house under certain building restrictions are typical of the way simple geometry can be used to solve problems.

Unit 10C explores a few basic ideas underlying the relatively new and exciting area of mathematics called fractal geometry. You may have heard about fractals and even seen one (the front cover of your text is decorated with a fractal), but now you will know how to make your own, among other things.

Unit 10A Fundamentals of Geometry

Overview

Because much of this chapter is based on ideas and concepts from geometry, it makes sense to do some review. If much of this unit is familiar, just sit back and enjoy it.

The unit begins by defining and giving examples of the basic concepts of geometry: **points**, **lines**, **planes**, and **angles**. We then move on to familiar

two-dimensional objects (also called plane objects). The most basic objects are **circles** and **regular polygons** such as **triangles**, **squares**, **pentagons**, and **hexagons**. You should know or be able to find quickly formulas for the
- area of a circle,
- circumference of a circle,
- area of a triangle, and
- area of a **parallelogram** (which include squares and rectangles).

We give several practical examples of the uses of these formulas.

The next subject is solid objects in three dimensions. You should be familiar with the table that has formulas for the
- surface area and volume of a **rectangular prism** (box),
- surface area and volume of a **cylinder** (soda can),
- surface area and volume of a **sphere**.

We give several examples of these formulas applied to practical problems.

The unit closes with a very useful and far-reaching section on scaling laws. We first discuss scale models, such as maps or architectural models. If we take an object and increase all of its dimensions by a factor of, say 10, then the **scale factor** is 10. The important result of this section is that
- areas scale with the *square* of the scale factor, and
- volumes scale with the *cube* the scale factor.

Thus, if you were to enlarge yourself by a scale factor of two, your surface area would increase by a factor of $2^2 = 4$, and your volume (and weight) would increase by a factor of $2^3 = 8$.

These scaling laws explain a lot of interesting phenomena. By considering the **surface-area-to-volume-ratio**, we can answer questions such as, why does crushed ice keep your drink colder than large ice cubes? Why can flies walk on ceilings? Why does the Moon have no active volcanoes?

Key Words and Phrases

geometry	Euclidean geometry	point
line	plane	dimension
angle	vertex	right angle
straight angle	acute angle	obtuse angle
perpendicular	parallel	radius
diameter	polygon	regular polygon
circumference	perimeter	parallelogram
rectangular prism	cube	cylinder
sphere	scale factor	scaling laws
surface to volume ratio		

Key Concepts and Skills

- define and give examples of point, line, and plane.
- convert angle measurements to fractions of a circle and vice versa.
- determine the perimeter and area of common plane objects (circle, triangle, square, rectangle, parallelogram).
- determine the surface area and volume of common three-dimensional objects (cube, rectangular prism, cylinder, sphere).
- use geometrical ideas and formulas to solve practical problems.
- understand scaling laws and surface-area-to-volume ratios.

Unit 10B Problem Solving with Geometry

Overview

Much of this chapter is based on ideas and concepts from geometry, but it involves applications which may be new to you. The unit begins by defining and giving examples of the basic concept of angles and how to measure them. The first application uses the perhaps familiar notion of **latitude** and **longitude**. As you may recall, this way of measuring the earth gives coordinates for locations with respect to the **prime meridian** and the **equator**. Knowing latitude and longitude, one can easily calculate the distance between two locations on the same meridian.

We explore the relationship between **angular size, physical size** and **distance** to an object. Using this relationship, one can calculate any one of the three from the other two. For example, one can estimate the diameter of the moon (its physical size) knowing the distance to it and its angular size.

The next application involves the notions of **pitch**, **grade**, and **slope.** They are basically the same thing, but one is usually preferred in a given context.

No discussion of geometry and measurement is complete without mentioning the **Pythagorean Theorem** and **similar triangles**. The first gives the familiar relationship between the lengths of the sides in a right triangle: $a^2 + b^2 = c^2$. The second gives a relationship between the sides of two triangles whose angles are all equal. This relationship is $\dfrac{a'}{a} = \dfrac{b'}{b} = \dfrac{c'}{c}$ where a, b, and c are the lengths of the sides in one triangle corresponding to the lengths a', b', and c' of the similar triangle. A nice application of similar right triangles is the calculation of heights from ground measurements such as the solar access problem of example 8.

The last section has to do with **optimal shapes**. As it turns out

∞ squares are optimal rectangles in that they maximize area while minimizing perimeter.

∞ cubes are optimal boxes in that they maximize volume while minimizing surface area.

Key Words and Phrases

latitude	longitude	prime meridian
equator	angular size	physical size
pitch	grade	slope
Pythagorean Theorem	similar triangles	optimal shape

Key Concepts and Skills

∞ use latitude to find distance between two locations with same longitude.

∞ use any two of angular size, physical size, and distance to an object to find the third.

∞ find the slope from pitch or grade and/or find a distance or height from them.

∞ find a distance using Pythagorean Theorem; also calculate the area of a right triangle.

∞ identify similar triangles and use them to solve measurement problems.

Unit 10C Fractal Geometry

Overview

In this unit we explore perhaps the most recent mathematical development in this book. **Fractal geometry** arises with the observation that classical geometry (invented by the ancient Greeks and studied by all of us in high school) works very well for regular objects such as circles and squares. But many objects in the real world, particularly natural forms, are far more complex and are not described well by classical geometry. Fractal geometry was proposed, in part, to describe the complicated forms we find in nature.

We begin by looking at what happens when we measure the length of a line segment. If we continually magnify the line we don't see anything new. This means that if we measure the length of a line segment, we get the same length regardless of the length of the ruler. Similarly, if we measure the area of a square, we get the same area regardless of how small the ruler might be. And if we measure the volume of a cube, we get the same volume regardless of how small the ruler is. This leads to the conclusion that a line is a one-dimensional object, a square is a two-dimensional object, and a cube is a three-dimensional object.

Here is how we define the **dimension** of an object. We imagine successively reducing the length of our ruler by a **reduction factor** R. Each time we reduce the length of the ruler, we observe by what factor N the number of **elements** increases. In the case of regular objects, we find that…

- for a one-dimensional object (e.g., a line segment), $N = R^1$.
- for a two-dimensional object (e.g., a square), $N = R^2$.
- for a three-dimensional object (e.g., a cube), $N = R^3$.

For irregular objects, the result may be quite different. If we carry out the same measurement process, we may find that the relationship between R and N has the form $N = R^D$, but D is no longer a whole number. If D is not a whole number, the object is called a **fractal**.

With this new definition for the dimension of a fractal object in hand, we look at some examples of fractals, notably the **snowflake curve,** the **Cantor set,** the **Sierpinski triangle**, and the **Sierpinski sponge**. We also discuss how to measure the fractal dimension of realistic objects such as coastlines. We close with the most famous fractal of all, the **Mandelbrot set**.

Key Words and Phrases

fractal geometry	element	reduction factor
fractal dimension	snowflake curve	snowflake island
self-similar	Sierpinski triangle	Sierpinski sponge
Cantor set	Mandelbrot set	

Key Concepts and Skills

- understand the meaning of dimension for regular objects (line, square, and cube).
- understand the meaning of dimension for fractal objects.
- determine the dimension of simple fractal objects.

11   Mathematics and the Arts

Overview

In this chapter we explore some very different applications of mathematics, namely music and the fine arts. As you will see, the connections between mathematics and the arts go back to antiquity. We devote one unit to music, one unit to classical painting, and the last to proportion as it appears in art and nature. This chapter should provide you with a new perspective and change of pace.

Unit 11A Mathematics and Music

Overview

The ties between mathematics and music go back to the ancient Greeks (~500 B.C.). The followers of Pythagoras discovered some of the basic laws that underlie our understanding of music today. They realized that the **pitch** of a musical note created by a plucked string is determined by the length of the string. This led to the more modern notions of the string's **fundamental frequency** (which measures how many times a string vibrates each second), and the principle that when the length of a string is halved, its pitch doubles (and goes up by an **octave**). With these few facts we can explain a lot.

In a standard scale that you might play on a piano or a guitar, one octave consists of 12 tones or **half-steps** (for example, the white and black keys between middle C and the next higher C). Here is the basic question we address: If we know the frequency of the first note of the scale, can we find the frequency of all 12 notes of the scale?

We show that to move up the scale a half step, we must *multiply* the frequency of the current tone by a fixed number. We give a brief argument showing that the magic number that generates the entire scale by multiplication is $f = 1.055946 \ldots$ or the twelfth root of 2. It turns out the notes of the scale follow an exponential growth law, as discussed in Chapter 9.

When two notes, played together, have a pleasing sound (**consonant tones**), the ratio of their frequencies is found to be (nearly) the ratio of two small numbers, such as 3/2 or 4/3. We investigate how these ratios compare to the exact frequencies generated by the magic number f, and discuss the problem of *temperament*.

The unit closes with observations about the modern connections between music and mathematics: digital music, compact disks, and synthesizers.

Key Words and Phrases

sound wave	pitch	frequency
cycles per second	fundamental frequency	overtones
octave	scale	half-step
consonant tones	analog	digital

Key Concepts and Skills

- understand the relation between frequency and pitch.
- determine the frequency of notes separated by an octave.
- determine the frequency of notes on a 12-tone scale, given the frequency of the first note.
- understand the difference between analog and digital music.

Unit 11B Perspective and Symmetry

Overview

It wasn't until the Renaissance (14th and 15th century) that painters attempted to draw three-dimensional objects realistically on a flat two-dimensional canvas. It took many years for these painters to perfect this technique, but in the end, they made a science of **perspective** drawing. In this unit we trace the development of perspective drawing and explore some of its mathematical necessities.

The key concept in perspective drawing is the **principal vanishing point**. It can be summarized as follows: If you are an artist standing behind a canvas, painting a real scene, then all lines that are parallel in the real scene and perpendicular to the canvas, must meet in a single point in the painting — this is the principal vanishing point. All other sets of parallel lines in the real scene (that are *not* perpendicular to the canvas) meet in their own vanishing points. All of the vanishing points (principal and otherwise) lie along a single line called the **horizon line**.

Another fundamental property of paintings and other objects of art is **symmetry**. Symmetry can mean many different things, but it often refers to a sense of balance. We define three different kinds of symmetry:

- **reflection symmetry**: an object can be "flipped" across a particular line and it remains unchanged (for example, the letter H).
- **rotation symmetry**: an object can be rotated through a particular angle and it remains unchanged (for example, the letter O).
- **translation symmetry**: an object or a pattern can be shifted, say to the right or the left, and it remains unchanged (for example, the pattern ....XXXXXX.... extended in both directions).

We investigate these symmetries in both geometrical objects and in actual paintings.

Symmetry arises in beautiful ways in **tilings** — patterns in which one or a few simple objects are used in repetition to fill a region of the plane. We give several examples of how triangles and quadrilaterals can be translated and reflected to produce wonderful patterns. You'll get the opportunity to try some tilings in the exercises.

Key Words and Phrases

vanishing point	principal vanishing point	horizon line
symmetry	reflection symmetry	rotation symmetry
translation symmetry	tiling	

Key Concepts and Skills

- understand the role of the principal vanishing point in perspective drawing.
- draw simple objects in perspective using vanishing points.
- identify symmetries in simple objects.
- draw simple objects with given symmetries.
- create tilings from triangles or quadrilaterals using translations and reflections.

Unit 11C Proportion and the Golden Ratio

Overview

In this unit, we explore another fundamental aspect of art: proportion. Issues of proportion arise not only in the art created by humans, but in natural forms as well. The subject has a lot to do with aesthetics, our innate sense of what is beautiful. One of earliest statements about proportion and beauty goes back to the ancient Greeks, who defined and studied the **golden ratio**.

The first instance of the golden ratio arises in dividing a line segment. What division of a line segment has the most visual appeal and balance? The Greeks argued that the most visually appealing division is the one where

$$\frac{\text{length of long piece}}{\text{length of short piece}} = \frac{\text{length of entire piece}}{\text{length of long piece}},$$

that is, the ratio of the length of the long piece to the length of the short piece is the same as the ratio of the length of the entire segment to the length of the long piece. We can also write this as

$$\frac{L}{1} = \frac{L+1}{L}.$$

The value of L that solves this relationship is the special (irrational) number

$$\phi = \frac{1+\sqrt{5}}{2} = 1.61803$$

This number is called the **golden ratio** or **golden section**.

From the golden ratio, we can define a **golden rectangle**. Any rectangle with sides in the ratio of ϕ is called a golden rectangle. Both the golden ratio and the golden rectangle arise throughout the history of art. Great architectural works (for example, the Greek Parthenon) have dimensions close to those of the golden rectangle (although some claim that these examples are coincidences). Many common objects such as post cards and cereal boxes have dimensions of golden rectangles. We also look at some examples of how the golden rectangle appears in nature.

There is one final connection that is too intriguing to ignore. We introduce the **Fibonacci sequence**, which was first used as a population model in the 13th century. Each term of the sequence is formed by adding the two previous terms:

$$1, 1, 2, 3, 5, 8, 13, 21, 34, \ldots .$$

We show how this sequence is related, perhaps unexpectedly, to the golden ratio. And the circle is closed by observing that the Fibonacci sequence also appears in the artwork of nature. This is a unit with several seemingly unrelated ideas that eventually become linked in a beautiful way.

Key Words and Phrases

proportion	golden ratio	golden rectangle
logarithmic spiral	symmetry	Fibonacci sequence

Key Concepts and Skills

- understand the golden ratio as a proportion and divide a line segment according to the golden ratio.
- construct and identify golden rectangles.
- generate the Fibonacci sequence and find the ratio of successive terms.
- understand the connection between the golden ratio and the Fibonacci sequence.

12   Mathematics and Politics

Overview

This chapter explores the surprisingly crucial role that mathematics plays in political matters. The first two units deal with elections. In elections with more than two candidates, there are several ways to determine a winner, and they may not agree. In fact, it can be shown that there is no single voting method that meets certain fairness criteria. Unit 12C is devoted to another political process, namely apportionment. How do we determine the number of representatives that each state sends to Congress? Again, there are several methods that can be used, and they do not always agree. Finally, having discussed the problems of apportionment, the last unit discusses the contentious issue of redistricting.

Unit 12A Voting: Does the Majority Always Rule?

Overview

In this and the following unit we investigate mathematical problems associated with voting. This may sound like an unusual application of mathematics, but voting problems have been studied for several centuries and it has long been known that curious things can happen in voting systems.

The discussion begins with elections between two candidates. Throughout these units a candidate can be interpreted as a choice between two or more alternatives. For example, a candidate may be a person running for office or a brand of bagels in a taste test. With only two candidates, the outcome is straightforward: the **majority** rules. This means that the candidate with the most votes (which must be more than 50% of the vote) wins the election.

However, even with majority rule, there are some interesting situations that can arise. We first look at presidential elections in which the winner is chosen, not by the **popular vote**, but by the **electoral vote**. Historically, there have been U.S. presidential elections in which a candidate won the popular vote, but lost the election.

We then look at variations on majority rule that involve **super majorities**. For example, many votes in the U.S. government require more than a 50% majority: it takes a 2/3 super majority in both houses of Congress, followed by a 3/4 super majority vote of the states to amend the U.S. Constitution. More than a 50% majority of a jury is required to reach a verdict in a criminal trial (a unanimous decision is often required).

Things get interesting when we turn to elections with three or more candidates. Often such elections are based on a **preference schedule** in which each voter ranks the candidates in order of preference. For example, the following preference schedule shows the outcome of an election among five candidates that we call A, B, C, D, and E.

First	A	B	C	D	E	E
Second	D	E	B	C	B	C
Third	E	D	E	E	D	D
Fourth	C	C	D	B	C	B
Fifth	B	A	A	A	A	A
	18	12	10	9	4	2

There were a total of 55 voters and 18 voters ranked A first, D second, E third, C fourth, and B last. The other columns are interpreted in a similar way. The question is: how do we determine a winner of the election?

The remainder of the unit presents five methods to determine a winner to an election with a preference schedule.

- **plurality**: the candidate with the most first-place votes wins (candidate A would win the above election).
- **top-two runoff**: the top two candidates have a runoff in which the votes of the losing candidates are redistributed to the top two candidates (candidate B would win the above election).
- **sequential runoff**: the candidates with the fewest first place votes are successively eliminated one at a time, votes are redistributed, and runoff elections are held at each stage (candidate C would win the above election).
- **point system** (or **Borda count**): with five candidates, five points are awarded for each first place vote, four points are awarded for each second place vote, and so on. The candidate with the most points wins (candidate D would win the above election).
- **pairwise comparison**: the winner between each pair of candidates is determined and the candidate with the most pairwise wins is the winner of the election (candidate E would win the above election).

You can probably already see the dilemma. We have proposed five reasonable methods for finding a winner and they all give different results for the above preference schedule. The unit closes inconclusively with this question unanswered. The next unit takes up the issue of fairness in voting systems and attempts to resolve the question.

Key Words and Phrases

majority rule	popular vote	electoral vote
super majority	preference schedule	plurality
single runoff	sequential runoff	point system
Borda count	pairwise comparisons	filibuster

Key Concepts and Skills

- understand the concept of majority rule and apply it in two-candidate elections.
- know the difference between popular vote and electoral vote.
- use super majority rules to determine the outcome of votes.
- create and interpret preference schedules.
- apply the methods of plurality, single runoff, sequential runoff, point system, and pairwise comparisons to determine the outcome of elections with preference schedules.

Unit 12B Theory of Voting

Overview

The previous unit closed with the observation that five reasonable methods for determining the winner of an election with three or more candidates can lead to five different winners. This unit attempts to resolve the dilemma, but as you will see, the resolution may be less than satisfying.

The analysis of voting methods involves what are known as **fairness criteria**. We will work with the following four criteria.

- **Criterion 1 (majority criterion):** If a candidate receives a majority of the first-place votes, that candidate should be the winner.
- **Criterion 2 (Condorcet criterion):** If a candidate is favored over every other candidate in pairwise races, then that candidate should be declared a winner.
- **Criterion 3 (monotonicity criterion):** Suppose that Candidate X is declared the winner of an election, and then a second election is held. If some voters rank X higher in the second election (without changing the order of other candidates), then X should also win the second election.
- **Criterion 4 (independence of irrelevant alternatives criterion):** Suppose that Candidate X is declared the winner of an election, and then a second election is held. If voters do not change their preferences, but one or more of the losing candidates drops out, then X should also win the second election.

These criteria have been developed by researchers who study voting methods; they are reasonable conditions that we would expect of any fair voting system. The first criterion, stating that a candidate with a majority of the votes should win, is the most natural. The second criterion is also straightforward: if a candidate has more votes than each of the other candidates taken individually, then that candidate should win the election. The third criterion just says that if some voters were to change their votes in favor of the winner of the election, then that candidate should still win the election. The last criterion says that if a losing candidate drops out of the election, the original winner should remain the winner.

Most of the unit is devoted to examining the five voting methods introduced in the previous unit to see whether they satisfy the four fairness criteria. You may be surprised at the results: none of the five methods always satisfies all of the fairness criteria. This is the message of a famous result in voting theory called **Arrow's Impossibility Theorem**. It says that there is no voting method that satisfies the four fairness criteria for all preference schedules.

We next look at a method called **approval voting** that has been proposed as an alternative way to handle elections with many candidates. It departs from the familiar one-person-one-vote idea that underlies all of the voting systems discussed so far (and it, too, has its drawbacks).

The unit closes with some ideas about power blocks in voting. An example of this idea is the formation of **coalitions,** which is common not only in our own government, but even more so in foreign governments.

Key Words and Phrases

fairness criteria   Arrow's Impossibility Theorem   approval voting

voting power   coalition

Key Concepts and Skills

- understand the four fairness criteria.
- apply the four fairness criteria to each of the five voting methods (introduced in UNIT 12A) for a given preference schedule.
- understand the benefits and disadvantages of approval voting.
- analyze an election in terms of coalitions.

Unit 12C Apportionment: The House of Representatives and Beyond

Overview

The U.S. Constitution stipulates that each state should have representation in the House of Representatives proportional to its population. But it doesn't specify exactly how the number of representatives should be determined. The process of assigning representatives according to population, which has applications beyond the House of Representatives, is called **apportionment**.

Apportionment begins by computing a **standard divisor**, which is the average number of people in the population per representative. For example, if the total population is 100,000 and there are 100 representatives, then the standard divisor is 1000 people per representative. The next step is to compute the **standard quota** for each state, which is the number of representatives a state should have if fractional representatives were possible. For example, if a state has 5500 people, then with a standard divisor of 1000, that state should have $5500/1000 = 5.5$ representatives. Now the dilemma of apportionment can be seen: it is not possible to have fractions of representatives.

The founding fathers proposed several methods to overcome this problem of fractional representatives. Here is a quick survey of the various methods that have been used over the years (and those considered in detail in the text).

- **Hamilton's method**: First, round the standard quota for each state *down* to form the **minimum quota** for that state; then give any remaining representatives to the states with the largest fraction of representatives.

- **Jefferson's method**: Choose a **modified divisor** such that when the new standard quotas are rounded *down* (to form a **modified quota**), all the representatives are used.
- **Webster's method**: Choose a **modified divisor** such that when the new standard quotas are rounded *according to the standard rounding rule* (to form a **modified quota**), all the representatives are used.
- **Hill-Huntington method**: Choose a **modified divisor** such that when the new standard quotas are rounded *according to a modified rounding rule* (to form a **modified quota**), all the representatives are used.

Notice that the methods appear quite similar, but involve slightly different rules for rounding. However, as the examples in the text show, the methods can produce different results. Many examples of are provided to show how these methods are carried out.

Which method is best? Two hundred years of American history have proved that none of the methods is perfect. Each method can exhibit at least one of several paradoxes or can violate certain reasonable expectations. To address this question of fairness, the unit explores each of the following situations:

- the **Alabama paradox**: when the total number of representatives increases, but at least one state actually loses representatives.
- the **population paradox**: when the population increases and under reapportionment a slow-growing state gains representatives at the expense of a fast-growing state.
- the **new states paradox**: when additional representatives are added to accommodate a new state and the apportionment of existing states changes.
- the **quota criterion**: under any apportionment, the number of representatives for any state should be one of the whole numbers nearest its standard quota (the standard quota rounded up or rounded down).

The apportionment problem is a bit like the voting problem, in that a perfect apportionment method cannot be found. This is the conclusion of the **Balinsky and Young theorem**.

Key Words and Phrases

apportionment	standard divisor	standard quota
Hamilton's method	Alabama paradox	population paradox
new states paradox	Jefferson's method	modified divisor
modified quota	quota criterion	Webster's method
Hill-Huntington method	geometric mean	Balinsky and Young theorem

Key Concepts and Skills

- understand the problem and the dilemma of apportionment.

- apply the four methods of the text to apportionment problems.
- identify the deficiencies of the four methods in terms of paradoxes and the quota criterion.

Unit 12D Dividing the Political Pie

Overview

The difficulties of apportionment, discussed in Unit 12C, do not stop after it is determined how many House seats each state is to acquire. Within each state, a political battle is waged every ten years (after each census) to decide how to divide the state into its congressional districts. This process, called **redistricting**, is most easily described as a mapping problem, where each state must redraw previous district lines to reflect shifts in the population, the goal being to divide the state into *contiguous* regions (reflecting the number of representatives) of nearly equal population.

In most states, redistricting decisions are made in the legislature, and this is where politics and mathematics meet. When a particular party has a majority in the state legislature, it often uses its political clout to draw lines that result in favorable outcomes for that party. Only a loose set of rules, set by recent U.S. Supreme court decisions, must be followed when redistricting, and thus politicians go to great lengths to create districts that will win their party as many House seats as possible. The practice of drawing district boundaries for political advantage is called **gerrymandering**.

The text provides several simplified examples of how political wrangling of this sort can very much influence both the number of seats a particular party wins, and the competitiveness of an election in a given district. Indeed, a striking graphic in Figure 12.4 clearly shows a reduction in competitive seats in the House over the last four decades.

The unit closes with ideas for reforming the redistricting system. Ultimately, the solution may lie in the hands of mathematicians, who could be called upon to provide a fair algorithm for redistricting.

Key Words and Phrases

    redistricting   gerrymandering   contiguous

Key Concepts and Skills

- understand the political and mathematical nature of redistricting.
- be aware of how redistricting affects the competitiveness of an election.
- understand the principals behind gerrymandering.
- apply the legal requirements for redistricting: equal populations in each district, and contiguous districts.

13   Mathematics and Business

Overview

Discrete mathematics is the area of mathematical inquiry that deals with separate or individual items. The items might be people, objects, choices, or geographical locations. We have already encountered some examples of discrete mathematics in this book: the counting methods presented in Unit 7E being a good case in point. This chapter will look at two more topics from discrete mathematics.

In Units 13A and B, we will look at applications of *network theory*. A network is a mathematical model that illustrates how the various components of a system are interrelated. For example, the power lines that connect several communities or the trading relations between several countries can be displayed using a network. The applications of network theory are endless. The mathematics is quite different because it is visual and involves very little computation. In the last unit, we see how networks can be used to schedule large projects that involve many stages. These methods are used to achieve efficiency and cost-savings in many business applications.

Unit 13A Network Analysis

Overview

In this unit we introduce the idea of a **network** (also called a **graph**). A network is a collection of points (called **vertices**) connected by lines or **edges**. A nice example of a network arises in the 18th century **Konigsberg bridge problem** that led to the invention of network theory. We also give additional situations that lead to networks.

One of the fundamental network questions is whether it is possible to find a path that traverses every edge of the network exactly once and returns to the starting vertex. If such a path exists, it is called an **Euler circuit** (pronounced *oiler*, in honor of the mathematician who invented network theory). An Euler circuit would be of interest to an inspector, a meter reader, or a delivery person, who would like to do his job in the most efficient way. It turns out there is a simple rule that determines whether a network has an Euler circuit:

> an Euler circuit exists for a network if and only if each vertex has an even number of edges attached to it.

Having determined whether a network has an Euler circuit doesn't tell you how to find it. So we also provide an algorithm, called the **burning bridges rule**. The rule states:

you may begin your circuit from any vertex in the network. However, as you choose edges to follow for your path, avoid using an edge that is the *only* connection to a part of the network that you have not already visited.

There is a lot of terminology associated with networks, some of which we introduce at this point. Here are a few important terms.

- a path within a network that begins and ends at the same vertex is a **circuit**.

- a network is **complete** if every vertex is connected to every other vertex.

- a **tree** is a network in which all of the vertices are connected, but there are no cycles (it looks like a tree with roots and branches).

- the **order** of a network is the number of vertices.

- the **degree** of a vertex is the number of edges connected to it (each vertex in a network could have a different order).

The last problem in this unit deals with networks in which every edge has a number or **weight** associated with it. The weights often represent distances or costs. A **spanning network** for such a network is a set of edges within the network that connects every vertex to every other vertex. In other words, if you walk along a spanning network, you can reach every vertex from every other vertex. A given network could have several different spanning networks, so a question of great importance arises: which, of many different spanning networks, has the minimum cost? This is called the **minimum cost spanning network** problem. (You might want to convince yourself that the minimum cost spanning network must be a tree).

We give some practical examples of minimum cost spanning network problems and then provide an efficient method, called **Kruskal's algorithm**, for finding the minimum cost spanning tree of a network.

Key Words and Phrases

network	edge	vertex
Euler circuit	burning bridges rule	complete network
circuit	tree	order
degree	spanning network	Kruskal's algorithm
minimum cost spanning network		

Key Concepts and Skills

- draw a network as a model of a given practical situation.
- determine if a given network has an Euler circuit and find one.
- understand the applications of Euler circuits.

- classify a network, and identify the order of the network and the degree of each vertex.
- understand the applications of minimum cost spanning networks.
- apply Kruskal's algorithm to find the minimum cost spanning network.

Unit 13B The Traveling Salesman Problem

Overview

This unit is devoted to one of the most famous of all network problems. However the **traveling salesman problem** has many applications beyond its namesake. Consider a network in which the vertices might be the cities that a salesman must visit and the edges are the airline routes that connect them along with their distances. A very practical problem is to find the route of minimum length that visits every city exactly once and returns to the starting city. This is an example of a traveling salesman problem.

We begin the discussion by introducing a preliminary idea: a **Hamiltonian circuit** is a path through a network that visits every vertex exactly once and returns to the starting vertex. While a Hamiltonian circuit passes through every vertex, it may not traverse every edge (as an Euler circuit does). Not every network has a Hamiltonian circuit, and furthermore, it is very difficult to determine whether a general network has a Hamiltonian circuit. In fact, there is no simple rule (as there is with Euler circuits) that determines when a network has a Hamiltonian circuit.

There are a few cases in which it is possible to find Hamiltonian circuits easily. If a network is complete (every vertex is connected to every other vertex), then Hamiltonian circuits abound. In fact there are $(n - 1)!/2$ Hamiltonian circuits in a complete network of order n. The number of Hamiltonian circuits increases explosively with the order of the network.

We must now imagine a network (which may or may not be complete) in which all of the edges have distances on them. To solve the traveling salesman problem for this network, we must find not just any Hamiltonian circuit, but the one with the least length. This turns out to be an incredibly difficult challenge, especially for problems of practical interest in which there may be thousands of vertices in the network. We describe a method, called the **nearest neighbor method**, which can be used to find nearly optimal solutions to the traveling salesman problem.

We close the unit with some general remarks about a branch of mathematics called **operations research** that includes problems such as the traveling salesman problem. Operations research often handles problems from business and management that involve finding the most efficient or least expensive way to carry out a task. Airlines and manufacturing companies use operations research every day to streamline their operations, and our lives are affected by this area of mathematics continuously.

Key Words and Phrases

traveling salesman problem

Hamiltonian circuit

nearest neighbor method

operations research

Key Concepts and Skills

- understand the applications of Hamiltonian circuits.
- determine if a given circuit is a Hamiltonian circuit for a network.
- find the number of Hamiltonian circuits in a complete network.
- understand the applications of the traveling salesman problem.
- apply the nearest neighbor method to find a traveling salesman solution.

Unit 13C Scheduling Problems

Overview

This unit introduces another practical use of networks. Imagine a large project, such as building a house or planning a shopping center (or even landing a person on the Moon), that consists of many smaller tasks. Some of the smaller tasks can be done at the same time, but some must be done before or after others. For example, in a house-building project, you must build the walls before you can run the electrical wiring within them.

A network can be used to schedule the smaller tasks and illustrate which tasks must precede other tasks. The network also shows the estimated time for the completion of each task. Once the scheduling network is created, several important questions can be answered. For example, we can determine the minimum time for completion of the project. The minimum completion time corresponds to the *longest* path through the network; this longest path is called the **critical path**. The critical path consists of all the **limiting tasks** for the project.

We can also determine when each task can be started and finished. The following four quantities are useful for this purpose:

- **earliest start time** (EST) for a task is the soonest that a task can be started after the beginning of the project.
- **earliest finish time** (EFT) for a task is the soonest that a task can be completed after the beginning of the project.
- **latest start time** (LST) for a task is the latest time at which it can start without delaying the overall project.
- **latest finish time** (LFT) for a task is the time by which it must be completed to avoid delays.

A general rule is that you should start at the *beginning* of the project and work *forward* to find the EST and EFT for various tasks. Start at the *end* of the project and consider each task in reverse order to find the LST and LFT.

Finally, for each task we define the **slack time** as the difference between the latest start time and the earliest start time (or equivalently as the difference between the latest finish time and the earliest finish time). Tasks with a zero slack time are on the critical path. Tasks with a non-zero slack time are not on the critical path and can be delayed without affecting the schedule of the entire project. Knowing the slack times allows a project manager to decide which tasks must be completed on time and which tasks can be postponed.

Key Words and Phrases

critical path	limiting task	earliest start time
earliest finish time	latest start time	latest finish time
slack time		

Key Concepts and Skills

- create and interpret a network as a schedule for a multi-stage project.
- find the critical path for a scheduling network.
- find the earliest start time, earliest finish time, latest start time, latest finish time, and slack times for all tasks in a scheduling network.

UNIT 1A

QUICK QUIZ

1. **a**. By the definition used in this book, an argument always contains at least one premise and a conclusion.
3. **b**. An argument must contain a conclusion.
5. **b**. Using the fact that a statement is unproved to imply that it is false is appeal to ignorance.
7. **b**. This is a personal attack because the premise (I don't trust his motives) attacks the character of the President, and says nothing about the substance of his tax plan.
9. **b**. Just because *A* preceded *B* does not necessarily imply that *A* caused *B*.

DOES IT MAKE SENSE?

15. Does not make sense. Raising one's voice has nothing to do with logical arguments.
17. Makes sense. A logical person would not put much faith in an argument that uses premises he believes to be false to support a conclusion.
19. Does not make sense. One can disagree with the conclusion of a well-stated argument regardless of whether it is fallacious.

BASIC SKILLS AND CONCEPTS

21. a. *Premise*: 70% of the TV audience watched the last round of *American Idol*. *Conclusion*: It must be worth watching.
    b. The fact that a lot of people watched *American Idol* does not imply it is worthwhile viewing.
23. a. *Premise*: No proof exists that global warming will have bad consequences. *Conclusion*: There's no reason to be concerned about global warming.
    b. The fact that something is not proven does not imply it is untrue.
25. a. *Premise*: He refused to testify. *Conclusion*: He must be guilty.
    b. There are many reasons that someone might have for refusing to testify (being guilty is only one of them), and thus this is the fallacy of limited choice.
27. a. *Premise*: He has an alleged history of borrowing others' work. *Conclusion*: I doubt he wrote that poem.
    b. The speaker is attacking the author's character to reach his conclusion, rather than providing evidence about the poem in question.
29. a. *Premise*: Crime has been decreasing in New York City. *Conclusion*: We should not build more prisons.

b. Attention from the real issue of building prisons (presumably in areas that could be far removed from New York) is being diverted by noting that crime in New York City is decreasing. Also, crime rates might not have much to do with the need for more prisons – there may be more important factors, such as the willingness of law enforcement to carry out its job.

31. *Premises*: President Reagan built up the defense system. Afterward, the Soviet Union began to break up. *Conclusion*: Reagan is responsible for the changes leading to the demise of the Soviet Union. This argument suffers from the **false cause** fallacy. It's true that Reagan's defense buildup preceded the demise of the Soviet Union, but we cannot conclude that the former caused the latter simply because it happened first.
33. The question asked by the lawyer assumes this *premise*: The mother is now using or has used drugs in the past. *Conclusion*: She should not be granted custody of her children. **Limited choice** is used in this argument, as the question does not allow for the possibility that the mother does not use drugs.
35. *Premises*: My mom smoked when she was my age. *Conclusion*: I'm not going to pay attention to her request that I not smoke. This is a **personal attack** on the mother's past transgressions, which should play little part in the child's logical decision about whether to smoke.
37. *Premise*: The ivory-billed woodpecker hasn't been seen in 50 years. *Conclusion*: It is extinct. **Appeal to ignorance** is used here – the lack of proof of the existence of the woodpecker does not imply it is extinct.
39. *Premise*: Gonzales belongs to the Sierra Club. *Conclusion*: He will not support additional logging in the national forests. This is a **personal attack**, on both Gonzales' membership in the Sierra Club, and the policies of the club, neither of which will necessarily compel Gonzales to vote one way or the other.
41. *Premises*: The percentage of those over 18 who smoke has decreased from 40% of the population to 20%. During the same time, the percentage of those who are overweight has increased from 25% to 35%. *Conclusion*: Quitting smoking leads to overeating. This is **false cause**: the fact that the percentage of smokers has decreased while the percentage of overweight people has increased does not prove the conclusion. (Furthermore, we are not told whether the overweight people are in the group of smokers).
43. *Premise*: My boy loves dolls, and my girl loves trucks. *Conclusion*: There's no truth to the claim that boys prefer mechanical toys while girls prefer

maternal toys. Using one child of each gender to come up with a conclusion about all children is **hasty generalization**. It can also be seen as an **appeal to ignorance**: the lack of examples of boys enjoying mechanical toys (and girls maternal toys) does not mean that they don't enjoy these toys.

45. *Premise*: The Republicans favor repealing the estate tax, which falls most heavily on the wealthy. *Conclusion*: Republicans think the wealthy aren't rich enough. (Implied here is that you should vote for Democrats). The argument distorts the position of the Republicans (none of whom would say the rich need to be richer); this is a **straw man**.

UNIT 1B

QUICK QUIZ

1. **c**. This is a proposition because it is a complete sentence making a claim, which could be true or false.
3. **c**. Conditional statements are, by definition, in the form of *if p, then q*.
5. **c**. Because it is not stated otherwise, we are dealing with the inclusive *or* (and thus either *p* is true, or *q* is true, or both are true).
7. **c**. The inclusive disjunction *Calcutta or Bombay* allows for either or both (i.e. at least one) of these words to appear in the article.
9. **b**. Statements are logically equivalent only when they have the same truth values.

DOES IT MAKE SENSE?

17. Does not make sense. Propositions are never questions.
19. Makes sense. If restated in *if p, then q* form, this statement would read, "If we catch him, then he will be dead or alive." Clearly this is true, as it covers all the possibilities. (One could argue semantics, and say that a dead person is not caught, but rather discovered. Splitting hairs like this might lead one to claim the statement does not make sense).
21. Does not make sense. Not all statements fall under the purview of logical analysis.

BASIC SKILLS AND CONCEPTS

23. This is not a proposition because it is not a complete sentence, nor does it make a claim to which one could assign a truth value.
25. This is a proposition as we can assign a truth value to it, and it's a complete sentence.

27. This is a complete sentence that makes a claim (a false one in this case), so it's a proposition.
29. Paris is not the capital of Maine. The original statement is false, so its negation is true.
31. The Nile River is in Africa. The original proposition is false, so its negation is true.
33. The city council supports the police chief.
35. Congress decided to honor the veto of the tax increase bill, which means Congress did not approve the bill.
37. The Constitution allows for the anti-discrimination policy.
39. This is the truth table for the conjunction *q and r*.

q	r	q and r
T	T	T
T	F	F
F	T	F
F	F	F

41. "Boston is the capital of Massachusetts" is false. "Moscow is the capital of Russia" is true. The conjunction is false because both propositions in a conjunction must be true for the entire statement to be true.
43. The two propositions are "Ben is married," and "Ben is a bachelor." We can't determine the truth value for either statement, but the conjunction must be false, because both statements can't be true at the same time.
45. "Cats have four legs" is true (in general), as is "Whales have no legs," so the conjunction is true.
47. This is the truth table for *q and r and s*.

q	r	s	q and r and s
T	T	T	T
T	T	F	F
T	F	T	F
T	F	F	F
F	T	T	F
F	T	F	F
F	F	T	F
F	F	F	F

49. *Or* is used in the exclusive sense because you probably can't have both the salad and dessert.
51. The exclusive *or* is used here as it is unlikely that the statement means you might do both tonight.
53. This could be thought of in both ways. If the exclusive *or* is used, the shorts will be made purely of cotton or wool. The inclusive *or* allows for a cotton/wool blend.

55. This is the truth table for the disjunction *r or s*.

r	s	r or s
T	T	T
T	F	T
F	T	T
F	F	F

57. This is the truth table for *p and (not p)*.

p	not p	p and (not p)
T	F	F
F	T	F

59. This is the truth table for *p or q or r*.

p	q	r	p or q or r
T	T	T	T
T	T	F	T
T	F	T	T
T	F	F	T
F	T	T	T
F	T	F	T
F	F	T	T
F	F	F	F

61. "The Amazon River is in South America" is true. "Beijing is in Germany" is false. The disjunction is true because a disjunction is true when at least one of its propositions is true.

63. The two propositions are "The light is on" and "The light is off." We can't determine the truth value for either proposition, though one of them must be true, so the disjunction is true.

65. "Beethoven was a professional ball player" is false. "Gandhi was a professional ball player" is also false. Since both are false, the disjunction is false.

67. Use option **a**, and search for *poverty* AND *Miami*.

69. Use options **a** and **c**, and search for (*museums* OR *concerts*) AND *San Francisco*.

71. This is the truth table for *if p, then r*.

p	r	if p, then r
T	T	T
T	F	F
F	T	T
F	F	T

73. *Hypothesis*: Trout can swim. *Conclusion*: Trout are fish. Since both are true, the implication is true, because implications are always true except in the case where the hypothesis is true and the conclusion is false.

75. *Hypothesis*: Italy is in Europe. *Conclusion*: China is in Africa. Since the hypothesis is true, and the conclusion is false, the implication is false (this is the only instance when a simple *if p, then q* statement is false).

77. *Hypothesis*: $4 \times 4 = 10$. *Conclusion*: America has a queen. Since the hypothesis is false, the implication is true, no matter the truth value of the conclusion (which, in this case, is false).

79. *Hypothesis*: $2 + 2 = 5$. *Conclusion*: Automobiles can swim. Since the hypothesis is false, the implication is true, no matter the truth value of the conclusion (which, in this case, is false).

81. If a person is a resident of St. Paul (*p*), then that person is a resident of Minnesota (*q*).

83. If you are considered a writer (*p*), then you are a novelist (*q*).

85. If a person has prostate cancer (*p*), then that person is a male (*q*).

87. *Converse*: If Anne lives in New Orleans, then she lives in Louisiana. *Inverse*: If Anne doesn't live in Louisiana, then she doesn't live in New Orleans. *Contrapositive*: If Anne doesn't live in New Orleans, then she doesn't live in Louisiana. The converse and inverse are always logically equivalent, and the contrapositive is always logically equivalent to the original statement.

89. *Converse*: If my heart rate increases, then I am running. *Inverse*: If I don't run, then my heart rate won't increase. *Contrapositive*: If my heart rate does not increase, then I am not running. The converse and inverse are always logically equivalent, and the contrapositive is always logically equivalent to the original statement.

91. *Converse*: If there's gas in the tank, the car's engine is running. *Inverse*: If the car's engine isn't running, then there is no gas in the tank. *Contrapositive*: If there is no gas in the tank, then the car's engine isn't running. The converse and inverse are always logically equivalent, and the contrapositive is always logically equivalent to the original statement.

93. If a young man has not wept, then he is a savage. If an old man will not laugh, then he is a fool. Opinions will vary on whether these are true.

95. If a person has no vices, then that person has very few virtues. Opinions will vary on whether this is true.

97. "If Sue lives in Cleveland, then she lives in Ohio," where it is assumed that Sue lives in Cincinnati. (Answers will vary). Because Sue lives in Cincinnati, the hypothesis is false, while the conclusion is true, and this means the implication is true. The converse, "If Sue lives in Ohio, then

she lives in Cleveland," is false, because the hypothesis is true, but the conclusion is false.

99. If Washington, D.C. is the capital of the United States, then Germany is in Asia. (Answers will vary). The proposition is false, because the hypothesis is true while the conclusion is false. The converse, "If Germany is in Asia, then Washington, D.C. is the capital of the United States," is true, as is any conditional proposition with false hypothesis.

101. a. The book would be on the list because the search finds all entries with either *Steinbeck* or *grapes* (or both).
b. The book would be on the list because the search finds all entries with *Grapes of Wrath*.
c. The book would be on the list because the search finds all entries with both *grapes* and *Steinbeck*.
d. The book would be on the list because the search would find all entries with *Steinbeck* (it would also find entries with *Hemingway*).
e. The book would not be on the list because the search would find only those items that have both *Hemingway* and *grapes*.
f. The book would be on the list because the search would find all entries with *Steinbeck* and *grapes* (it would also find all items with *Hemingway* and *grapes*).

103. "If a computer exhibits human-level intelligence, then it is capable of massive processing power, around 20 million billion calculations per second." The above statement converts the *necessary* part of the quote into an *if...then* statement. Kurzweil is claiming that any computer capable of mimicking human intelligence must also be capable of performing 20 million billion calculations per second (presumably because our brains possess such processing power). Note that Kurzweil states that massive processing power is not a *sufficient* condition. If it were, he'd be claiming, "If a computer is capable of massive processing power, then it exhibits human-level intelligence." Thus Kurzweil recognizes that even if a computer could be designed to perform 20 million billion calculations per second, it may not be capable of exhibiting human-level intelligence.

105. Following is a truth table for both *not (p or q)* and *(not p) and (not q)*.

p	q	p or q	not (p or q)	(not p) and (not q)
T	T	T	F	F
T	F	T	F	F
F	T	T	F	F
F	F	F	T	T

Since both statements have the same truth values (compare the last two columns in the table), they are logically equivalent.

107. Following is a truth table for *not (p or q)* and *(not p) or (not q)*.

p	q	p or q	not (p or q)	(not p) or (not q)
T	T	T	F	F
T	F	T	F	T
F	T	T	F	T
F	F	F	T	T

Note that the last two columns in the truth table don't agree, and thus the statements are not logically equivalent.

109. Following is a truth table for *(p or q) and r* and *(p and r) or (q and r)*.

p	q	r	p or q	(p or q) and r	p and r	q and r	(p and r) or (q and r)
T	T	T	T	T	T	T	T
T	T	F	T	F	F	F	F
T	F	T	T	T	T	F	T
T	F	F	T	F	F	F	F
F	T	T	T	T	F	T	T
F	T	F	T	F	F	F	F
F	F	T	F	F	F	F	F
F	F	F	F	F	F	F	F

Since the fifth and eighth columns agree, the statements are logically equivalent.

UNIT 1C

QUICK QUIZ

1. **b.** The ellipsis is a convenient way to represent all the other states in the U.S. without having to write them all down.

3. **a.** When the circle labeled C is contained within the circle labeled D, it indicates that C is a subset of D.

5. **a.** Because all apples are fruit, the set A should be drawn within the set B (the set of apples is a subset of the set of fruits).

7. **a.** The X is placed in the region where *business executives* and *working mothers* overlap to indicate that there is at least one member in that region.

9. **a.** The central region is common to all three sets, and so represents those who are male, Republican, and an athlete.

DOES IT MAKE SENSE?

17. Does not make sense. More likely than not, the payments go to two separate companies.
19. Does not make sense. The number of students in a class is a whole number, and whole numbers are not in the set of irrational numbers.
21. Does not make sense. A Venn diagram shows only the relationship between members of sets, but does not have much to say about the truth value of a categorical proposition.

BASIC SKILLS & CONCEPTS

23. 23 is a natural number.
25. 2/3 is a rational number.
27. 1.2345 is a rational number.
29. π is a real number.
31. −34.45 is a rational number.
33. $\pi/4$ is a real number.
35. −13/3 is a rational number.
37. $\pi/129$ is a real number.
39. {1900s, 1910s, 1920s,…, 1990s}.
41. {Washington, Adams, Jefferson}.
43. {California, Colorado, Connecticut}.
45. {50, 51, 52,…,400}.
47. Because some women are veterinarians, the circles should overlap.

49. All verbs are words, and thus the set of verbs is a subset of the set of words. This means one circle should be contained within the other.

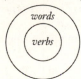

51. Some painters are also musicians, so the circles should overlap.

53. No negative integer is a natural number, so these sets are disjoint, and the circles should not overlap.

55. The subject is *widows*, and the predicate is *women*.

57. All U.S. presidents are people over 30 years old. The subject is *U.S. presidents*, and the predicate is *people over 30 years old*.

59. No monkey is a gambling animal. The subject is *monkeys*, and the predicate is *gambling animals*.

61. All nurses are people who know CPR. The subject is *nurses*, and the predicate is *people who know CPR*.

63.

65.

things that are not songs, novels or published works

unpublished songs that are novels (no members)

unpublished songs

unpublished novels

published songs that are novels (no members)

songs   *novels*

published songs

published works

published novels

published works that are neither songs nor novels

67.

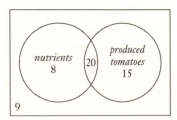

clothes that are hats

things that are not clothes, shirts, or hats

clothes other than hats and shirts

clothes   *hats*

hats that are not clothes (no members)

clothes that are hats and shirts (no members)

clothes that are shirts

shirts

hats that are shirts (no members)

shirts that are not clothes (no members)

69. a. There are 17 men in the room.
   b. There are 19 non-Republicans in the room.
   c. There are 6 Republican women in the room.
   d. There are 32 people in the room.

71.

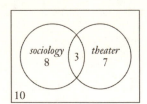

sociology 8   3   *theater* 7

10

73.

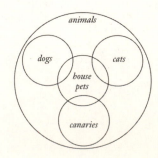

nutrients 8   20   *produced tomatoes* 15

9

FURTHER APPLICATIONS

75. 26 non-attorneys attended, and 17 of the non-Americans were not attorneys.

attorneys 45   23   *Americans* 9

17

77.

women 110   32   *biology* 21

87

79.

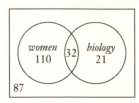

coffee 4806   91   *gall stones* 385

14,068

81. a.

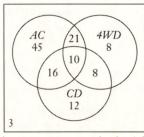

AC 45   21   *4WD* 8

10

16   8

CD 12

3

b. The region common to both AC and 4WD shows that 31 people chose both features (this includes those who also chose a CD player).
c. Count the number of people that are in the AC circle, but outside the 4WD circle, and add them together to arrive at 61 people.
d. All of the regions contained within the three circles correspond to those who chose at least one feature (which is how one can interpret AC or 4WD or CD). There are 120 such people.
e. Add the regions that are common to two circles, but not all three, to find that 45 people chose exactly two options.

83.

animals

dogs   *cats*

house pets

canaries

85.

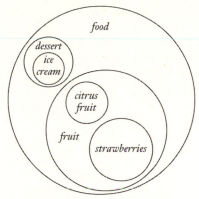

87. The two-way table should look like this:

	Showed improvement	No improvement	Total
New Drug	20	15	35
Placebo	25	40	65
Total	45	55	100

This is one possible Venn diagram.

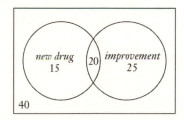

89. Two dogs are purebreds with long tails.

91. As shown in the diagram, the sets *hairy animals* and *fish* do not overlap (because all hairy animals are mammals, and no mammal is a fish). Since some mammals can swim, and hairy animals are mammals, there may be hairy animals that swim. There is nothing in the propositions that excludes the possibility of walking mammals, and thus there may be walking mammals, and for the same reason, there may be hairy animals that walk on land.

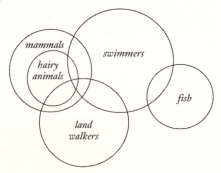

93. There could be conservative Democrats, and there could be liberal Green Party members, because there is nothing in the propositions to prohibit this. No liberal is a Republican, because the set of Republicans is a subset of the set of conservatives, and no liberal is a conservative (which means those sets are disjoint).

95. a.

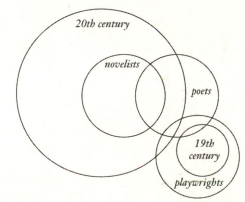

b. No, all of the novelists were born in the 20th century.

c. There is nothing in the facts you noticed that excludes the possibility of a poet born in the 19th century, so, yes, you may have studied such a poet.

d. Yes, there's nothing stated that excludes this possibility.

97. a. There are 16 different sets of options: choose nothing, A, B, C, D, AB, AC, AD, BC, BD, CD, ABC, ABD, ACD, BCD, and ABCD.

b. No, it is not possible (not with circles, in any case – one can draw three interlocking circles followed by a spiral-shaped tube that snakes through the requisite regions to show all the choices, but the diagram begins to get complicated to the point where it becomes less useful).

c. There are 32 different sets of options: choose nothing, A, B, C, D, E, AB, AC, AD, AE, BC, BD, BE, CD, CE, DE, ABC, ABD, ABE, ACD, ACE, ADE, BCD, BCE, BDE, CDE, ABCD, ABCE, ABDE, ACDE, BCDE, and ABCDE.

d. Notice that with four options, there are $2^4 = 16$ different sets of options, and that with five options, there are $2^5 = 32$ different sets of options. It turns out that this pattern continues so that with n options, there are 2^n different sets of options.

UNIT 1D

QUICK QUIZ

1. **b.** The only way to prove a statement true beyond all doubt is with a valid and sound deductive argument.
3. **c.** If a deductive argument is not valid, it cannot be sound.
5. **c.** Diagram *a* in question 4 is the correct diagram for its argument, and since X lies on the border of the *knights* circle, Paul may or may not be a knight.
7. **c.** This argument is of the form *affirming the conclusion*, and it is always invalid, which means we can conclude nothing about *p*.
9. **b.** The side opposite the right angle in a right triangle is always the longest, and it's called the hypotenuse.

DOES IT MAKE SENSE?

19. Does not make sense. One cannot prove a conclusion beyond all doubt with an inductive argument.
21. Makes sense. As long as the logic of a deductive argument is valid, if one accepts the truth (or soundness) of the premises, the conclusion necessarily follows.
23. Does not make sense. This argument is of the form *affirming the conclusion*, and it is always invalid.

BASIC SKILLS & CONCEPTS

25. This is an inductive argument because it makes the case for a general conclusion based on many specific observations.
27. This is an inductive argument because it makes the case for a general conclusion based on many specific observations.
29. This is a deductive argument because a specific conclusion is deduced from more general premises.
31. This is an inductive argument because it makes the case for a general conclusion based on several specific observations.
33. It is difficult to assert whether the premises are true because there are so many car models made by each of the manufacturers listed, with widely varying gas mileage, and the concept of good gas mileage is not well defined. If the premises are accepted to be true, the argument is moderately strong, and the conclusion may be true (especially

when comparing Japanese made cars to American made cars).

35. The premises may be true (they likely all wear expensive basketball shoes, in any case), but the argument is weak, and the conclusion is false.
37. The premises are true, and the argument is moderately strong. The conclusion could be true.
39. Premise: All islands are tropical lands. The diagram shows the argument is valid. However, it is not sound as the first premise is false.

41. Premise: Fast foods are foods that increase cholesterol. The argument is valid, and the premises are true, so the argument is also sound.

43. Premise: All states north of the Mason Dixon line are states that get snow. The argument is valid, and the premises are true, so the argument is also sound.

45. Premise: All opera singers are people who can whistle a Mozart tune. The diagram shows the argument is valid, and it is sound if we believe the premises to be true (there's room for disagreement on that count as we can't be sure all opera singers can whistle, nor can we be sure they know Mozart).

47. Affirming the hypothesis – this form is always valid, as confirmed by the diagram. The premises are true, and thus the argument sound.

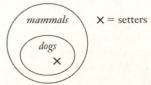

49. Affirming the hypothesis – this form is always valid, as confirmed by the diagram. As long as the premises are true, the argument is sound.

51. Denying the hypothesis – this form is always invalid, as confirmed by the diagram. Since it is invalid, it cannot be sound.

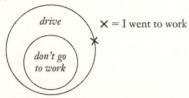

53. Affirming the conclusion – this form is always invalid, as confirmed by the diagram. Since it is invalid, it cannot be sound.

55. The argument is in standard form, and it is valid as there is a clear chain of implications from the first premise to the conclusion.

57. The second premise and conclusion should be written as follows to put the argument in standard form. Premise: If taxpayers have less disposable income, the economy will slow down. Conclusion: If taxes are increased, then the economy will slow down. The argument is valid as there is a clear chain of implications from premises to conclusion.

59. The statement is true.

61. The statement is not true (consider negative numbers, or just about any other pair of numbers).

FURTHER APPLICATIONS

63. Answers will vary. An example:
    Premise: All living mammals breathe.
    Premise: All monkeys are mammals.
    Conclusion: All living monkeys breathe.

65. Answers will vary. An example:
    Premise: All mammals fly. (false)
    Premise: All monkeys are mammals. (true)
    Conclusion: All monkeys fly. (false)

67. Answers will vary. An example:
    Premise: All mammals breathe. (true)
    Premise: All mammals have hair. (true)
    Conclusion: All hairy animals breathe. (true)

69. An example of *affirming the conclusion* (invalid):
    Premise: If I am in Phoenix, then I am in Arizona.
    Premise: I am in Arizona.
    Conclusion: I am in Phoenix.

71. An example of *denying the conclusion* (valid):
    Premise: If I am in Phoenix, then I am in Arizona.
    Premise: I am not in Arizona.
    Conclusion: I am not in Phoenix.

UNIT 1E

QUICK QUIZ

1. **b**. A vote for C implies a property tax reduction.

3. **c**. With unknown application fees, it's not clear which bank has the better offer.

5. **c**. $20/100 min = 20¢/min.

7. **c**. Both Jack's argument and this one make a huge leap from premises to conclusion: here's a few examples when *A* happened, so *A* must happen all the time.

9. **a**. The teacher is assuming that students will do fine without spell checkers, which implies that traditional methods of teaching spelling are effective.

DOES IT MAKE SENSE?

15. Makes sense. The double negative means the insurance company accepted his claim.

17. Does not make sense. If Sue wants to save time, she should take the Blue Shuttle, and save ten minutes.

19. Makes sense. Both the duration and mileage of the first warranty is the better deal.

BASIC SKILLS & CONCEPTS

21. He has $2 left as he gave away the rest.

23. No, it's not possible, as a man who has a widow is dead.

25. You must meet eight people, as the first six might be Democrats.

27. You must meet three people. The first and second might be from different parties, in which case by the third person, you will have met two from the same party.

29. Jane might go to the movies 2, 3, 4, 5, or 6 days per week.

31. No it does not follow. Some of the communications majors may be men.

33. a. The state constitution is less strict, as the initiative seeks to prevent state courts from interpreting the right of free expression more broadly under the state constitution than under the First Amendment.

b. A *yes* vote would allow for stronger laws against obscenity.

c. False. If the initiative passed, then state laws could be as strict as the First Amendment calls for, but not more strict.

35. (1) We have enough money to afford a house. (2) We will be staying in the same place long enough to make buying a house sensible. (Answers will vary).

37. (1) Not all countries get along well at present. (2) The United Nations can help nations learn to get along. (Answers will vary).

39. We are looking for possible unstated motives that may be the unstated "real reason" for opposition to the tax cuts. Among the possibilities: (1) a desire to maintain or increase government spending programs, which are more likely to be cut if revenues fall; (2) a belief that the tax system should redistribute wealth from the wealthy to the poor, which is less likely to occur with tax cuts.

41. In all of these problems, the key is to understand situation (iii). Focusing on Monica from part **c**, notice that her earned income plus $250 is $4150, and that number is "the greater of $700 or earned income plus $250" (i.e. $4150 > $700). Since her total of earned and unearned income is $4200, and this is more than the number arrived at using (iii) (i.e. $4200 > $4150), she must file a return. The other situations are similar: compute earned income (up to $4050) plus $250, compare it to $700, take the greater of those two numbers, and compare that result to earned plus unearned income to decide whether the person needs to file a return.

a. Trent does not need to file a return.

b. Sally does not need to file a return.

c. Monica must file a return.

d. Horace does not need to file a return.

e. Delila must file a return.

43. a. The landlord has one month after June 5 to return the deposit, so the terms have been met.

b. The landlord has one month after June 5 to return the deposit, so the terms have been met.

c. The landlord has one month after June 5 to return the deposit, so the terms have not been met.

45. You need estimates of the number of minutes you spend calling during weekday daytime, and during the weekend (the weekday nighttime minutes cost the same under both plans, and thus don't affect the comparison). If the number of weekday daytime minutes is less than one-third the number of weekend minutes, then the new plan is better.

FURTHER APPLICATIONS

47. Were there more than 350 people on each jet, or both jets combined? Was there any damage or injuries?

49. Is $83 billion the gross domestic product, or 6% of the gross domestic product?

51. a. They could be consistent, because Alice does not specify the time period for her 253 cases, and it's possible Zack is being selective about the cases he chooses to discuss (that is, he may not be talking about all the cases Alice tried).

b. They could be consistent if Alice obtained many convictions through plea agreements without going to trial.

53. Alicia is a biology major from Alabama. Bob is a communications major from Connecticut. Carla is an art major from Arkansas.

55. Under your current policy (and over the span of a nine-month pregnancy), you'll spend $115 per month, plus $4000 for prenatal care and delivery, for a total of $5035. Under the upgraded policy, you'll spend $275 per month, for a total of $2475. Thus, considering only these costs (we aren't told, for example, what happens if the mother requires an extensive hospital stay due to a C-section or other complications arising from delivery, or what happens if the baby is born prematurely and requires neo-natal care), the upgraded policy is best.

57. If you fly nine times with Airline A, at a cost of $3150, your tenth flight will be free (because you will have flown 27,000 miles at that point). If you fly ten times with Airline B, your cost will be $3250. Thus if you plan to fly ten times (or a multiple of ten times), Airline A is cheaper than Airline B.

59. The pitching has been getting worse (answers will vary).

61. The soil or watering was different in the two flowerbeds (answers will vary).

63. The number of students increased each year.

65. Country Y has a high gun suicide rate.

UNIT 2A

QUICK QUIZ

1. **a.** Think of the unit *miles per hour*; the unit of *mile* is divided by the unit of *hour*.
3. **c.** When multiplying quantities that have units, the units are also multiplied, so $ft^2 \times ft = ft^3$.
5. **a.** Divide both sides of 1 L = 1.06 qt by 1 L.
7. **b.** $1.20 per euro means 1 euro = $1.20, which is more than $1.
9. **b.** Intuitively, it makes sense to divide a volume (with units of, say, ft^3) by a depth (with units of ft) to arrive at surface area (which has units of ft^2).

DOES IT MAKE SENSE?

17. Does not make sense. 35 miles is a distance, not a speed.
19. Does not make sense. Two ft^2 describes an area, not a volume.
21. Does not make sense. Using the familiar formula *distance = rate × time*, one can see that it would be necessary to divide the distance by the rate (or speed) in order to compute the time, not the other way around.

BASIC SKILLS & CONCEPTS

23. a. $\frac{3}{4} \times \frac{1}{2} = \frac{3 \cdot 1}{4 \cdot 2} = \frac{3}{8}$.

b. $\frac{2}{3} \times \frac{3}{5} = \frac{2 \cdot 3}{3 \cdot 5} = \frac{2}{5}$.

c. $\frac{1}{2} + \frac{3}{2} = \frac{1+3}{2} = \frac{4}{2} = 2$.

d. $\frac{2}{3} + \frac{1}{6} = \frac{4}{6} + \frac{1}{6} = \frac{4+1}{6} = \frac{5}{6}$.

e. $\frac{2}{3} \times \frac{1}{4} = \frac{2 \cdot 1}{3 \cdot 4} = \frac{2}{12} = \frac{1}{6}$.

f. $\frac{1}{4} + \frac{3}{8} = \frac{2}{8} + \frac{3}{8} = \frac{2+3}{8} = \frac{5}{8}$.

g. $\frac{5}{8} - \frac{1}{4} = \frac{5}{8} - \frac{2}{8} = \frac{5-2}{8} = \frac{3}{8}$.

h. $\frac{3}{2} \times \frac{2}{3} = \frac{3 \cdot 2}{2 \cdot 3} = 1$.

25. Answers may vary depending on whether fractions are reduced.

a. $3.5 = \frac{35}{10} = \frac{7}{2}$. b. $0.3 = \frac{3}{10}$.

c. $0.05 = \frac{5}{100} = \frac{1}{20}$. d. $4.1 = \frac{41}{10}$.

e. $2.15 = \frac{215}{100} = \frac{43}{20}$. f. $0.35 = \frac{35}{100} = \frac{7}{20}$.

g. $0.98 = \frac{98}{100} = \frac{49}{50}$. h. $4.01 = \frac{401}{100}$.

27. a. 1/4 = 0.25. b. 3/8 = 0.375.
c. 2/3 = 0.667. d. 3/5 = 0.6.
e. 13/2 = 6.5. f. 23/6 = 3.833.
g. 103/50 = 2.06. h. 42/26 = 1.615.

29. The price of apple juice has units of dollars per ounces, or $/oz.
31. The cost of carpet has units of dollars per square yard, or $/ yd^2.
33. The price of eggplants has units of euros per kilogram, or euro/kg.
35. The gas mileage has units of miles per gallon, or mpg (or mi/gal).

37. $3.5 \text{ mi} \times \frac{5280 \text{ ft}}{1 \text{ mi}} = 18,480 \text{ ft}$.

39. $3.5 \text{ hr} \times \frac{60 \text{ min}}{1 \text{ hr}} = 210 \text{ min}$.

41. $3 \text{ lb} \times \frac{16 \text{ oz}}{1 \text{ lb}} = 48 \text{ oz}$.

43. $2 \text{ wk} \times \frac{7 \text{ day}}{1 \text{ wk}} \times \frac{24 \text{ hr}}{1 \text{ day}} \times \frac{60 \text{ min}}{1 \text{ hr}} = 20,160 \text{ min}$.

45. $35 \text{ gal} \times \frac{4 \text{ qt}}{1 \text{ gal}} = 140 \text{ qt}$.

47. a. The area of the reservoir's base is 120 ft × 40 ft = 4800 ft^2, and the amount of water the reservoir can hold is 120 ft × 40 ft × 9 ft = 43,200 ft^3.
b. The area of the soccer field is 110 yd × 55 yd = 6050 yd^2, and the volume of the snow is 110 yd × 55 yd × 3 ft = 110 yd × 55 yd × 1 yd = 6050 yd^3 (note that 3 ft was converted to 1 yd before computing the volume).
c. The area of the base is 120 ft × 120 ft = 14,400 ft^2, and the volume of the building is 120 ft × 120 ft × 330 ft = 4,752,000 ft^3.

49. Note that 1 ft = 12 in., and thus $(1 \text{ ft})^2 = (12 \text{ in.})^2$, which means 1 ft^2 = 144 $in.^2$. This can also be written as $\frac{1 \text{ ft}^3}{1728 \text{ in.}^3} = 1$, or $\frac{144 \text{ in}^3}{1 \text{ ft}^3} = 1$.

51. Since 1 m = 100 cm, we have $(1 \text{ m})^2 = (100 \text{ cm})^2$, which means 1 m^2 = 10,000 cm^2. This can also be written as $\frac{1 \text{ m}^2}{10,000 \text{ cm}^2} = 1$, or $\frac{10,000 \text{ cm}^2}{1 \text{ m}^2} = 1$.

53. The volume of the box is 12 ft × 8 ft × 4 ft = 384 ft^3. Since 1 yd = 3 ft, we know 1 yd^3 = 27 ft^3, and this can be used to convert to cubic yards.

$$384 \text{ ft}^3 \times \frac{1 \text{ yd}^3}{27 \text{ ft}^3} = 14.2 \text{ yd}^3.$$

55. Use the fact that 1 yd^3 = 27 ft^3 (see exercise 53).

$$320 \text{ ft}^3 \times \frac{1 \text{ yd}^3}{27 \text{ ft}^3} = 11.9 \text{ yd}^3.$$

57. Since $1 U.S. is worth 10.95 pesos, the dollar is worth more than the peso (which is about 9¢).

59. $2500 \text{ peso} \times \dfrac{\$1}{10.95 \text{ peso}} = \$228.31$, so 2500 pesos is worth 228.31 U.S. dollars. Note that if you use the conversion 1 peso = $0.09136, you'll get an answer of $228.40 U.S. (this is due to errors in rounding).

61. $12{,}000 \text{ yen} \times \dfrac{\$1}{111.5 \text{ yen}} = \$107.62$, so you can buy 107.62 U.S. dollars with 12,000 yen. Note that if you use the conversion 1 yen = $0.008965, you'll get an answer of $107.58 U.S. (this is due to errors in rounding).

63. $\dfrac{11 \text{ peso}}{1 \text{ L}} \times \dfrac{\$0.09136}{1 \text{ peso}} \times \dfrac{3.785 \text{ L}}{1 \text{ gal}} = \dfrac{\$3.80}{\text{gal}}$, so the price is $3.80 per gallon.

65. The car is traveling with speed $\dfrac{13 \text{ mi}}{15 \text{ min}}$, and this needs to be converted to mi/hr.

$$\frac{13 \text{ mi}}{15 \text{ min}} \times \frac{60 \text{ min}}{1 \text{ hr}} = 52 \frac{\text{mi}}{\text{hr}}.$$

67. Convert $436 million to people using the conversion 1 person = $8.

$$\$436{,}000{,}000 \times \frac{1 \text{ person}}{\$8} = 54{,}500{,}000 \text{ people},$$

so 54.5 million people saw the film.

69. Convert 4.7 pounds to dollars by using the conversion 1 lb = $1.29.

$$4.7 \text{ lb} \times \frac{\$1.29}{\text{lb}} = \$6.06.$$

71. First, note that 300 million people is 3000 groups of 100,000 people each. The mortality rate is

$$\frac{555{,}000 \text{ deaths}}{300{,}000{,}000 \text{ people}} \times \frac{300{,}000{,}000 \text{ people}}{3000 \text{ groups of } 100{,}000} = 185 \frac{\text{deaths}}{100{,}000}.$$

73. Convert 3 million births per year into births per minute.

$$\frac{3{,}000{,}000 \text{ births}}{\text{yr}} \times \frac{1 \text{ yr}}{365 \text{ d}} \times \frac{1 \text{ d}}{24 \text{ hr}} \times \frac{1 \text{ hr}}{60 \text{ min}} = 5.7 \frac{\text{births}}{\text{deaths}}.$$

75. The average speed in miles per hour is

$$\frac{1200 \text{ mi}}{24 \text{ hr}} = 50 \frac{\text{mi}}{\text{hr}}.$$

77. Convert one year into hours of sleep.

$$1 \text{ yr} \times \frac{365 \text{ d}}{\text{yr}} \times \frac{7 \text{ hr (of sleep)}}{\text{d}} = 2555 \text{ hr}.$$

79. The solution is wrong. It's helpful to include units on your numerical values. This would have signaled your error, as witness:

$$(\textit{incorrect}) \; 0.11 \text{ lb} \div \frac{\$7.70}{\text{lb}} = \frac{0.11 \text{ lb}^2}{\$7.70} = 0.014 \frac{\text{lb}^2}{\$}.$$

Notice that your solution, when the correct units are added, produces units of square pounds per dollar, which isn't very helpful, and doesn't answer the question. The correct solution is

$$0.11 \text{ lb} \times \frac{\$7.70}{\text{lb}} = \$0.85.$$

81. The solution is wrong. Units should be included with all quantities. When dividing by $11, the unit of dollars goes with the 11 into the denominator (as shown below). Also, while it's reasonable to round an answer to the nearest tenth, it's more useful to round to the nearest hundredth in a problem like this, as you'll be comparing the price per pound of the large bag to the price per pound of the small bag, which is 39¢ per pound. Here's your solution with all units attached, division treated as it should be, and rounded to the hundredth-place:

$$50 \text{ lb} \div \$11 = \frac{50 \text{ lb}}{\$11} = 4.55 \frac{\text{lb}}{\$}.$$

This actually produces some useful information, as you can see a dollar buys 4.55 pounds of potatoes. If you're good with numbers, you can already see this is a better buy than 39¢ per pound, which is roughly 3 pounds for a dollar. But it's better to find the price per pound for the large bag:

$$\$11 \div 50 \text{ lb} = \frac{\$11}{50 \text{ lb}} = 0.22 \frac{\$}{\text{lb}} = 22¢ \text{ per pound}.$$

Now you can compare it to the price per pound of the small bag (39¢/lb) and tell which is the better buy.

83. Convert 2500 miles into gallons by using the conversion 28 miles = 1 gallon.

$$2500 \text{ mi} \times \frac{1 \text{ gal}}{28 \text{ mi}} = 89.3 \text{ gal}.$$

85. a. The driving time when traveling at 55 miles per hour is $2000 \text{ mi} \times \frac{1 \text{ hr}}{55 \text{ mi}} = 36.4 \text{ hr}$, while the time at 70 miles per hour is $2000 \text{ mi} \times \frac{1 \text{ hr}}{70 \text{ mi}} = 28.6 \text{ hr}$.

b. Your car gets 38 miles to the gallon when driving at 55 mph, so the cost is

$$2000 \text{ mi} \times \frac{1 \text{ gal}}{38 \text{ mi}} \times \frac{\$3.25}{\text{gal}} = \$171.05.$$

Your car gets 32 miles to the gallon when driving at 70 mph, so the cost is

$$2000 \text{ mi} \times \frac{1 \text{ gal}}{32 \text{ mi}} \times \frac{\$3.25}{\text{gal}} = \$203.13 \ .$$

87. Visualize a rectangular slab of water, 75 feet long, 54 feet wide, and 6 inches high – this is the amount of water that needs to be added to the pool to raise the water level 6 inches. Since 6 in. = 0.5 ft, the volume of this slab is

$$75 \text{ ft} \times 54 \text{ ft} \times 0.5 \text{ ft} = 2025 \text{ ft}^3 \ .$$

FURTHER APPLICATIONS

89. The maximum house footprint allowed is 1/4 of a 1/4–acre lot, which is 1/16 of an acre. Convert this to square feet:

$$\frac{1}{16} \text{acre} \times \frac{43,560 \text{ ft}^2}{1 \text{ acre}} = 2722.5 \text{ ft}^2 \ .$$

91. Convert 2 gigabytes into pages, using 2000 bytes = 1 page.

$$2 \text{ Gbyte} \times \frac{1 \text{ billion bytes}}{\text{Gbyte}} \times \frac{1 \text{ page}}{2000 \text{ bytes}} = 1 \text{ million pages}$$

93. a. The volume of the bath is 6 ft × 3 ft × 2.5 ft = 45 ft^3. Fill it to the halfway point, and you'll use 22.5 ft^3 of water (half of 45 is 22.5). (Interesting side note: it doesn't matter which of the bathtub's three dimensions you regard as the height – fill it halfway, and it's always 22.5 ft^3 of water). When you take a shower, you use

$$1 \text{ shower} \times \frac{10 \text{ min}}{\text{shower}} \times \frac{1.75 \text{ gal}}{\text{min}} \times \frac{1 \text{ ft}^3}{7.5 \text{ gal}} = 2.33 \text{ ft}^3$$

of water, and thus you use considerably more water when taking a bath.

b. Convert 22.5 ft^3 of water (the water used in a bath) into minutes.

$$22.5 \text{ ft}^3 \times \frac{7.5 \text{ gal}}{1 \text{ ft}^3} \times \frac{1 \text{ min}}{1.75 \text{ gal}} = 96 \text{ min} \ .$$

c. Plug the drain in the bathtub, and mark the depth to which you would normally fill the tub when taking a bath. Take a shower, and note how long your shower lasts. Step out and towel off, but keep the shower running. When the water reaches your mark (you used a crayon, and not a pencil, right?), note the time it took to get there. You now have a sense of how many showers it takes to use the same amount of water as a bath. For example, suppose your shower took 12 minutes, and it takes a full hour (60 minutes) for the water to reach your mark. That would mean every bath uses as much water as five showers.

95. Convert 1 week into cubic feet of water.

$$1 \text{ wk} \times \frac{7 \text{ d}}{1 \text{ wk}} \times \frac{24 \text{ hr}}{1 \text{ d}} \times \frac{60 \text{ min}}{1 \text{ hr}} \times \frac{60 \text{ s}}{1 \text{ min}} \times \frac{25,800 \text{ ft}^3}{\text{s}}$$
$$= 15,603,840,000 \text{ ft}^3 \ ,$$

which is about 15.6 billion cubic feet. Since

$$\frac{15,603,840,000}{1,200,000,000,000} = 0.013 \ , \text{ about } 1.3\% \text{ of the water}$$

in the reservoir was released.

UNIT 2B

QUICK QUIZ

1. **c.** The metric prefix *kilo* means 1000, so a kilometer is 1000 meters.

3. **a.** A millimeter is one-thousandth of a meter (or 0.001 meter), which is about the size of the point on a pen.

5. **c.** One liter is equal to 1.057 quart, which is about one quart.

7. **b.** A watt is defined to be one joule per second, which is a unit of power, not energy.

9. **a.** Multiplying the population density, which has units of $\frac{\text{people}}{\text{mi}^2}$, by the area, which has units of mi^2, will result in an answer with units of people.

DOES IT MAKE SENSE?

19. Does not make sense. First of all, we use the unit of kilogram to measure mass, not weight, though mass and weight are often used interchangeably in everyday conversation. Even so, a bicyclist with mass of 300 kg would weigh more than 650 pounds (on the surface of the earth), which is an unheard of weight for a professional cyclist.

21. Does not make sense. The units of liters per second might be used to measure the flow rate of a faucet, but not the speed at which someone runs (meters per second would make more sense for that).

23. Makes sense. 10,000 meters is 10 kilometers, which is about 6.2 miles, a common length for foot races. Anyone who can run six back-to-back 9-minute miles has no trouble running 10,000 meters in less than an hour.

25. Does not make sense. The unit of meter measures length, not volume.

27. Does not make sense. Utility companies bill you for energy use (typically in units of kW-hr), not power ratings (recall that the watt is a unit of power).

BASIC SKILLS & CONCEPTS

29. $10^4 \times 10^7 = 10^{4+7} = 10^{11}$.

31. $10^6 \div 10^2 = 10^{6-2} = 10^4$.

33. $10^{-2} \times 10^{-6} = 10^{-2+(-6)} = 10^{-8}$.

35. $10^{12} \times 10^{23} = 10^{12+23} = 10^{35}$.

37. $10^{25} \div 10^{15} = 10^{25-15} = 10^{10}$.

39. $10^2 + 10^{-1} = 100 + 0.1 = 100.1$

41. $10^{12} \div 10^{-4} = 10^{12-(-4)} = 10^{16}$.

43. $10^4 + 10^2 = 10,000 + 100 = 10,100$.

45. Answers will vary. Consider a person who is 5 ft 10 in. tall. Since $5\text{ ft} = 5\text{ ft} \times \dfrac{12\text{ in.}}{1\text{ ft}} = 60\text{ in.}$, that person is 60 in. + 10 in. = 70 in. tall.

47. Convert a cubic foot of water into pounds.
$$1\text{ ft}^3 \text{ (of water)} \times \frac{7.48\text{ gal}}{1\text{ ft}^3} \times \frac{8.33\text{ lb}}{1\text{ gal}} = 62.3\text{ lb}.$$
Now convert that answer into ounces.
$$62.3\text{ lb} \times \frac{16\text{ oz}}{1\text{ lb}} = 997\text{ oz}.$$

49. Convert 2 gallons to pints, and then into cubic inches.
$$2\text{ gal} \times \frac{4\text{ qt}}{1\text{ gal}} \times \frac{2\text{ pt}}{1\text{ qt}} = 16\text{ pt};$$
$$16\text{ pt} \times \frac{28.88\text{ in.}^3}{1\text{ pt}} = 462\text{ in.}^3$$

51. Convert bushels to cubic feet.
$$100,000\text{ bushels} \times \frac{4\text{ pecks}}{1\text{ bushel}} \times \frac{8\text{ qt}}{1\text{ peck}} \times \frac{67.2\text{ in.}^3}{1\text{ qt}} \times \frac{1\text{ ft}^3}{1728\text{ in.}^3}$$
$$= 124,444\text{ ft}^3.$$

53. You can tell the factor by which the first unit is larger (or smaller) than the second by dividing one into the other. Rewrite metric prefixes as powers of 10, and then simplify, as shown below.
$$\frac{1\text{ cm}}{1\text{ km}} = \frac{10^{-2}\text{ m}}{10^3\text{ m}} = \frac{10^{-2}}{10^3} = 10^{-2-3} = 10^{-5},$$
which is one hundred-thousandth. This means a centimeter is 100,000 times smaller than a kilometer.

55. See exercise 53.
$$\frac{1\text{ L}}{1\text{ mL}} = \frac{1\text{ L}}{10^{-3}\text{ L}} = \frac{1}{10^{-3}} = 10^3 = 1000,$$ so a liter is 1000 times larger than a milliliter.

57. See exercise 53.
$$\frac{1\text{ m}^2}{1\text{ km}^2} = \frac{1\text{ m}^2}{(10^3\text{ m})^2} = \frac{1\text{ m}^2}{10^6\text{ m}^2} = \frac{1}{10^6} = 10^{-6}, \text{ which is}$$
one millionth, so a square meter is one million times smaller than a square kilometer.

59. $8\text{ m} \times \dfrac{3.28\text{ m}}{1\text{ ft}} = 26.2\text{ ft}$.

61. $10\text{ lb} \times \dfrac{1\text{ kg}}{2.205\text{ lb}} = 4.54\text{ kg}$.

63. $45\text{ L} \times \dfrac{1\text{ gal}}{3.785\text{ L}} = 11.89\text{ gal}$.

65. Cube both sides of the conversion 2.54 cm = 1 in. to find the conversion between cubic centimeters and cubic inches.
$$(2.54\text{ cm})^3 = (1\text{ in.})^3 \Rightarrow 16.387\text{ cm}^3 = 1\text{ in.}^3.$$
Now use this to complete the problem.
$$5\text{ cm}^3 \times \frac{1\text{ in.}^3}{16.387\text{ cm}^3} = 0.31\text{ in.}^3.$$

67. $\dfrac{10\text{ kg}}{1\text{ m}^3} \times \dfrac{2.205\text{ lb}}{1\text{ kg}} \times \left(\dfrac{1\text{ m}}{3.28\text{ ft}}\right)^3 = 0.62\dfrac{\text{lb}}{\text{ft}^3}$.

69. a. Use C = (F − 32)/1.8. C = (45 − 32)/1.8 = 7.2°C.
    b. Use F = 1.8C + 32. F = 1.8(20) + 32 = 68°F.
    c. F = 1.8(-15) + 32 = 5°F.
    d. F = 1.8(-30) + 32 = -22°F.
    e. C = (70 − 32)/1.8 = 21.1°C.

71. a. Use C = K − 273.15. C = 50 − 273.15 = -223.15°C.
    b. C = 240 − 273.15 = -33.15°C.
    c. Use K = C + 273.15. K = 10 + 273.15 = 283.15 K.

73. Power is the rate at which energy is used, so the power is 1000 Calories per hour. Convert this to joules per second, or watts.
$$\frac{1000\text{ Cal}}{1\text{ hr}} \times \frac{1\text{ hr}}{60\text{ min}} \times \frac{1\text{ min}}{60\text{ s}} \times \frac{4184\text{ j}}{1\text{ Cal}} = \frac{1162\text{ j}}{\text{s}} = 1162\text{ W}.$$
Yes, this is plenty to keep a 100-watt bulb shining, assuming the energy generated playing basketball can be converted to electrical energy without too much loss.

75. a. Convert kilowatt-hours into joules.
$$1000\text{ kW-hr} \times \frac{3,600,000\text{ j}}{1\text{ kW-hr}} = 3.6 \times 10^9\text{ j}.$$
   b. June has 30 days, so the average power is
$$\frac{3,600,000,000\text{ j}}{30\text{ d}} \times \frac{1\text{ d}}{24\text{ hr}} \times \frac{1\text{ hr}}{60\text{ min}} \times \frac{1\text{ min}}{60\text{ s}} = \frac{1389\text{ j}}{\text{s}} = 1389\text{ W}.$$
One could also begin with 1000 kW-hr per 30 days, and convert that into watts, using 1 watt = 1 joule per second.

   c. First, convert joules into liters.
$$3.6 \times 10^9\text{ j} \times \frac{1\text{ L}}{12,000,000\text{ j}} = 300\text{ L}.$$
Now convert liters into gallons.
$$300\text{ L} \times \frac{1\text{ gal}}{3.785\text{ L}} = 79\text{ gal}.$$

Thus it would take 300 liters = 79 gallons of oil to provide the energy shown on the bill (assuming all the energy released by the burning oil could be captured and delivered to your home with no loss).

77. Density is mass per unit volume, so the density of the pebble is $\frac{40 \text{ g}}{10 \text{ cm}^3} = 4 \frac{\text{g}}{\text{cm}^3}$. It will sink in water because the density of water is $1\frac{\text{g}}{\text{cm}^3}$.

79. Population density is people per unit area, so the average population density of the U.S. is
$$\frac{300,000,000 \text{ people}}{3,500,000 \text{ mi}^2} = 85.7 \frac{\text{people}}{\text{mi}^2}.$$

81. The population density for New Jersey is
$$\frac{8,400,000 \text{ people}}{7419 \text{ mi}^2} = 1132 \frac{\text{people}}{\text{mi}^2}.$$
The population density for Wyoming is
$$\frac{490,000 \text{ people}}{97,105 \text{ mi}^2} = 5 \frac{\text{people}}{\text{mi}^2},$$
which is smaller than New Jersey's population density.

83. a. BAC is usually measured in units of grams of alcohol per 100 milliliters of blood. A woman who drinks two glasses of wine, each with 20 grams of alcohol, has consumed 40 grams of alcohol. If she has 4000 milliliters of blood, her BAC is
$$\frac{40 \text{ g}}{4000 \text{ mL}} = \frac{0.01 \text{ g}}{\text{mL}} \times \frac{100}{100} = \frac{1 \text{ g}}{100 \text{ mL}}.$$
It is fortunate that alcohol is not absorbed immediately, because if it were, the woman would most likely die – a BAC above 0.4 g/mL is typically enough to induce coma or death.
b. If alcohol is eliminated from the body at a rate of 10 grams per hour, then after 3 hours, 30 grams would have been eliminated. This leaves 10 grams in the woman's system, which means her BAC is $\frac{10 \text{ g}}{4000 \text{ mL}} = \frac{0.0025 \text{ g}}{\text{mL}} \times \frac{100}{100} = \frac{0.25 \text{ g}}{100 \text{ mL}}$. This is well above the legal limit for driving, so it is not safe to drive. Of course this solution assumes the woman survives 3 hours of lethal levels of alcohol in her body, because we have assumed all the alcohol is absorbed immediately. In reality, the situation is somewhat more complicated.

85. a. A metric mile is $1500 \text{ m} \times \frac{3.28 \text{ ft}}{1 \text{ m}} = 4290 \text{ ft}$. A USCS mile is 5280 ft. Since 4290/5280 = 0.932, the metric mile is 93.2% of the USCS mile.

87. Mauna Kea is 18,200 ft + 13,796 ft = 31,996 ft when measured from its base on the ocean floor to its peak. This is $31,996 \text{ ft} \times \frac{1 \text{ mi}}{5280 \text{ ft}} = 6.06 \text{ mi}$,

and $31,996 \text{ ft} \times \frac{1 \text{ m}}{3.28 \text{ ft}} \times \frac{1 \text{ km}}{1000 \text{ m}} = 9.75 \text{ km}$. When measured this way, Mauna Kea is higher than Mt. Everest, but, in the opinion of this author, it would be silly to claim Mauna Kea as the highest mountain in the world. The altitude of a mountain is its distance above sea level, and it's hardly fair to use one definition of altitude for one mountain, and another definition for an island mountain. After all, one could argue that Mt. Everest rises from the Bay of Bengal in India, and since the bay extends deep into the Indian Ocean, Mt. Everest is significantly "taller" than the 29,035 ft used here.

89. 24-karat gold is 100% pure, and since 25% of 24 is $0.25 \times 24 = 6$, a nugget that is 25% pure is 6-karat gold.

91. $2.5 \text{ carats} \times \frac{0.2 \text{ g}}{1 \text{ carat}} = 0.5 \text{ g}$.

93. In order to compute the cost, we need to find the energy used by the refrigerator in kW-hr, and then convert to cents (after which it may make sense to convert to dollars). This can be done in a single chain of unit conversions, beginning with the idea that energy = power × time (this is true because power is the rate at which energy is consumed; that is, power = energy/time).
$$350 \text{ W} \times 1 \text{ yr} \times \frac{1 \text{ kW}}{1000 \text{ W}} \times \frac{365 \text{ d}}{1 \text{ yr}} \times \frac{24 \text{ hr}}{1 \text{ d}} \times \frac{10¢}{1 \text{ kW-hr}} \times \frac{\$1}{100¢}$$
= \$306.60, or about \$307.

95. Compute the energy used by a 75-watt bulb in 10,000 hours. This will be the energy saved when replacing a 100-watt bulb with a 25-watt bulb.
$$75 \text{ W} \times 10,000 \text{ hr} \times \frac{1 \text{ kW}}{1000 \text{ W}} = 750 \text{ kW-hr}.$$
Convert 750 kW-hr into cents (and then dollars) to find out how much money is saved.
$$750 \text{ kW-hr} \times \frac{10¢}{1 \text{ kW-hr}} \times \frac{\$1}{100¢} = \$75.$$

97. Since energy = power × time, the power plant can produce (assuming a 30-day month)
$$1.5 \times 10^9 \text{ W} \times 1 \text{ month} \times \frac{30 \text{ d}}{\text{month}} \times \frac{24 \text{ hr}}{\text{d}} \times \frac{1 \text{ kW}}{1000 \text{ W}}$$
$$= 1.08 \times 10^9 \text{ kW-hr} = 1.08 \text{ billion kW-hr}$$
of energy every month. The plant needs
$$1.08 \times 10^9 \text{ } \tfrac{\text{kW-hr}}{\text{month}} \times \frac{1 \text{ kg}}{450 \text{ kW-hr}} = 2.4 \times 10^6 \text{ } \tfrac{\text{kg}}{\text{month}},$$
or 2.4 million kg of coal every month. Because
$$1.08 \times 10^9 \text{ kW-hr} \times \frac{1 \text{ home}}{1000 \text{ kW-hr}} = 1.08 \times 10^6 \text{ homes},$$
the plant can serve 1.08 million homes.

99. At 20% efficiency, this solar panel can generate 200 watts of power when exposed to direct

sunlight. Since energy = power×time, and because the panel receives the equivalent of 6 hours of direct sunlight, the panel can produce

$$200 \text{ W} \times 6 \text{ hr} \times \frac{1 \text{ kW}}{1000 \text{ W}} \times \frac{3,600,000 \text{ j}}{\text{kW-hr}} = 4,320,000 \text{ j} .$$

This occurs over the span on one day, so the panel produces an average power of

$$\frac{4,320,000 \text{ j}}{\text{d}} \times \frac{1 \text{ d}}{24 \text{ hr}} \times \frac{1 \text{ hr}}{60 \text{ min}} \times \frac{1 \text{ min}}{60 \text{ s}} = 50 \frac{\text{j}}{\text{s}} = 50 \text{ W} .$$

101. Energy = power×time, so the energy produced by a wind turbine over the course of a year is

$$200 \text{ kW} \times 1 \text{ yr} \times \frac{365 \text{ d}}{1 \text{ yr}} \times \frac{24 \text{ hr}}{1 \text{ d}} = 1,752,000 \text{ kW-hr} .$$

This is enough energy to serve

$$1,752,000 \text{ kW-hr} \times \frac{1 \text{ home}}{10,000 \text{ kW-hr}} = 175 \text{ homes} .$$

UNIT 2C

QUICK QUIZ

1. **c.** Look at example 1 (*Box Office Receipts*) in this unit.

3. **b.** Common experience tells us that batteries can power a flashlight for many hours, not just a few minutes nor several years.

5. **b.** Refer to the discussion of Zeno's paradox in the text. There, it was shown that the sum of an infinite number of ever-smaller fraction is equal to 2, and thus we can eliminate answers **a** and **c**. This leaves **b**.

7. **c.** A widow is a woman who survives the death of her husband, and thus the man is dead, making it impossible for him to marry anyone.

9. **b.** The most likely explanation is that the A train always arrives 10 minutes after the B train. Suppose the A train arrives on the hour (12:00, 1:00, 2:00,…), while the B train arrives ten minutes before the hour (11:50, 12:50, 1:50,…). If Karen gets to the station in the first 50 minutes of the hour, she'll take the B train; otherwise she'll take the A train. Since an hour is 60 minutes long, 5/6 of the time, Karen will take the B train to the beach. Note that 5/6 of 30 days is 25 days, which is the number of times Karen went to the beach. The other scenarios *could* happen, but they aren't nearly as likely as the scenario in answer **b**.

DOES IT MAKE SENSE?

13. Does not make sense. There is no problem-solving recipe that can be applied to all problems.

15. Does not make sense. The four-step method is applicable to a wide variety of problems, and is not limited to mathematical problems. (It's not a cure-all, either, but it's quite useful, and worth learning).

BASIC SKILLS & CONCEPTS

17. You won't be able to determine the exact number of child and adult tickets sold, but the method of trial-and-error leads to the following possible solutions: (1 child, 6 adult), (4 child, 4 adult), (7 child, 2 adult), (10 child, 0 adult). Along the way, you may have noticed that the number of adult tickets must be an even number, because when it is odd, the remaining money cannot be divided evenly into $20 (child) tickets.

19. Jill runs 100 meters in 15 seconds, so she'll run 105 meters in 15.75 seconds. To see why this is so, treat her rate of 100 m/15 s as a unit conversion factor (the reciprocal of which can also be used): $105 \text{ m} \times \frac{15 \text{ s}}{100 \text{ m}} = 15.75 \text{ s}$. In the first race, Jack runs 95 meters in 15 seconds, so his time to the finish line (at 100 meters) in the second race is

$$100 \text{ m} \times \frac{15 \text{ s}}{95 \text{ m}} = 15.7895 \text{ s} .$$ The difference of these times, 15.7895 s – 15.75 s = 0.0395 s, is the amount of time by which Jill beats Jack.

21. Note that after two hours, the cars collide, because the first travels at 80 km/hr, and thus covers 160 km, while the second travels at 100 km/hr, and covers 200 km. The canary is flying at 120 km/hr for the entire 2 hours, so it has flown 240 km.

23. On the second transfer, there are four possibilities to consider. A) All three marbles are black. B) Two are black, one is white. C) One is black, two are white. D) All three are white. In case A), after the transfer, there are no black marbles in the white pile, and no white marbles in the black pile. In case B), one white marble is transferred to the black pile, and one black marble is left in the white pile. In case C), two white marbles are transferred to the black pile, and two black marbles are left in the white pile. In case D), three white marbles are transferred to the black pile, and three black marbles are left in the white pile. In all four cases, there are as many white marbles in the black pile as there are black marbles in the white pile after the second transfer.

25. Cut one of the columns lengthwise, and press it flat – this results in a rectangle with height = 12 feet, and width = circumference of column = 3 feet. As shown in example 5, a wrap of ribbon becomes the hypotenuse of a right triangle. The base of the triangle is 3 feet. Because there are ten wraps per column, its height is one-tenth the

height of the column, which is 12/10 = 6/5 ft. By the Pythagorean theorem, the length of one wrap of ribbon is $h = \sqrt{3^2 + \left(\frac{6}{5}\right)^2} = 3.231$ ft , and thus 32.31 feet of ribbon are required for one column (ten wraps, each of length 3.231 ft). To wrap all 40 columns, multiply 32.31 by 40 to arrive at 1292.4 feet.

27. If the rail forms a circular arc, as stated in the problem, the radius of this circle would be quite large, and the curvature of the arc would be very small. Thus a straight line is a reasonable approximation to the slowly-turning curve of the circular arc. A circle of radius 1/2-mile would not make a good approximation, because the length of the semi-circular arc so formed would be half the circumference of a circle of radius 1/2-mile. This comes out to 8294 ft, which is more than 3000 feet longer than the straight rail. However, we are told the straight rail only increases in length by one foot.

FURTHER APPLICATIONS

31. This problem can be solved by trial and error. Using one bag of topsoil, for example, leaves 22 cubic feet, which would require 11 bags of fertilizer. Thus one solution is (1 topsoil, 11 fertilizer). The others are found in a similar fashion: (3 topsoil, 8 fertilizer), (5 topsoil, 5 fertilizer), (7 topsoil, 2 fertilizer). You may have noticed along the way that there must be an odd number of bags of topsoil so that the remaining volume to be filled by fertilizer is divisible by two.

33. Following the hint given in the problem, the best place to start is with a drawing that shows the room cut along its vertical corners and laid flat. Figure 33a shows the room and the configuration of the amplifier and speaker before the cuts are made, and Figure 33b shows the room in "flat mode." Note that there are at least two solutions to consider (a third will be presented in Figure 33c).

FIGURE 33a

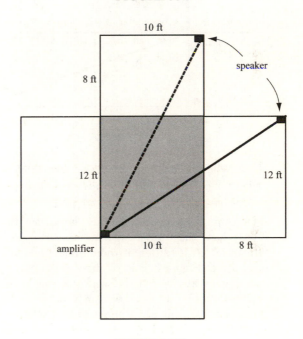

FIGURE 33b

In flat mode, it is clear that the shortest distance between the amplifier and the speaker is a straight line, though there are two possible paths (the upper right hand corner of the room, where the speaker is situated, must be represented twice, as shown in Figure 33b). In both cases, the distance from the amplifier to the speaker can be computed using the Pythagorean theorem.

Case 1 (lower triangle): the length of the wire is

$$\sqrt{18^2 + 12^2} = \sqrt{468} = 21.6 \text{ ft} .$$

Case 2 (upper triangle): the length of the wire is

$$\sqrt{20^2 + 10^2} = \sqrt{500} = 22.4 \text{ ft} .$$

Based on these calculations, the shortest length of wire needed to connect the amplifier to the speaker is 21.6 ft. However, there's a third possibility that is not evident when looking at Figure 33b. In order to see it, the room must be cut and laid flat in a different configuration, shown here.

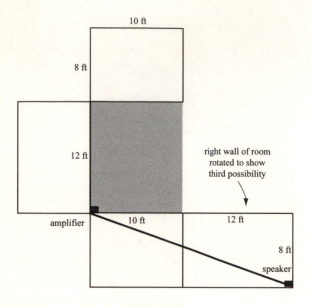

10 ft

8 ft

12 ft

amplifier

right wall of room rotated to show third possibility

10 ft   12 ft

8 ft

speaker

FIGURE 33c

This solution has the wire traveling along the walls of the room for its entire length, which is

$$\sqrt{22^2 + 8^2} = \sqrt{548} = 23.4 \text{ ft}.$$

Since this does not produce a shorter length of wire than the previous solution, the answer remains 21.6 ft (different room dimensions might reveal that Figure 33c is the best choice). There are other ways to run the wire not shown here, such as rotating the back wall in a manner similar to Figure 33c (this gives an equivalent solution of 23.4 ft), or running the wire along the ceiling. However, the problem states the wire is to run along the floor and walls, so the ceiling solutions need not be considered. Try some solutions along the ceiling if you're curious – you'll find they are equivalent to running the wire along the floor.

35. Assuming the endpoints of the bridge are at the level of the river, and that the length of the curved bridge can be reasonably approximated by two straight lines (as shown in Figure 35), the height of the bridge can be determined by the Pythagorean theorem.

distance along bridge = 202 m

h   101 m   100 m

river

200 m

FIGURE 35

$$h = \sqrt{101^2 - 100^2} = \sqrt{201} = 14 \text{ m}.$$

37. In going from the first floor to the third floor, you have gone up two floors. Since it takes 30 seconds to do that, your pace is 15 seconds per floor. Walking from the first floor to the sixth floor means walking up five floors, and at 15 seconds per floor, this will take 75 seconds.

39. You might pick one of each kind of apple on the first three draws. On the fourth draw, you are guaranteed to pick an apple that matches one of the first three, so four draws are required.

41. The blind fiddler is a woman.

43. Select a fruit from the box labeled *Apples and Oranges*. If it's an apple, that box must be the *Apple* box (because its original label is incorrect, leaving only *Apples* or *Oranges* as its correct label). The correct label for the box labeled *Oranges* is either *Apples* or *Apples and Oranges*, but we just determined the box containing apples, so the correct label for this box is *Apples and Oranges*. There's only one choice left for the box labeled *Apples*, and that's *Oranges*. A similar argument allows one to determine the correct labeling for each box if an orange is selected first.

45. Seven crossings are required. Trip 1: the woman crosses with the goose. Trip 2: the woman returns to the other shore by herself. Trip 3: the woman takes the mouse across. Trip 4: the woman returns with the goose. Trip 5: the woman brings the wolf across. Trip 6: the woman returns to the other shore by herself. Trip 7: the woman brings the goose over. In this way, the goose is never left alone with the wolf, nor the mouse with the goose. The foursome could get to the other side of the river in five crossings if you give rowing abilities to the animals, but the statement of the problem (and common sense) probably forbids it (the boat "…will hold only herself and one other animal"). Try it anyway for an interesting twist to the problem.

47. Ten rungs will be showing – the boat rises with the tide.

49. It's possible you'll pick a sock of each color on the first two draws, but on the third draw, you are assured of making a match.

51. The first prisoner would know he had a white hat if he saw red hats on prisoners 2 and 3. Since he doesn't know the color of his hat, prisoners 2 and 3 are both wearing white hats, or one has a white hat while the other has a red hat. If prisoner 2 saw a red hat on prisoner 3, he would know that he must have a white hat (he can't have a red hat while seeing red on prisoner 3, because then prisoner 1 would know his color). But he doesn't know, which means prisoner 3 must have a white hat. Prisoner 3 doesn't need to see the hats of the other prisoners; he can deduce that his hat is white simply by hearing the other two prisoners confess they don't know what hats they are wearing.

53. The coin has revolved twice (720° of revolution). The best way to see this is to duplicate the experiment. Another way to think about it: consider two circular clocks (instead of coins), nestled side by side, both in the same position, with 12-o'clock pointing straight ahead (see Figure 53). Roll the right clock around the left clock until the 12's are touching. The right clock has rotated 180°, but it's made only 1/4 of its journey around the left clock. By the time it makes the full journey around, it will have rotated 720°.

initial position of clocks

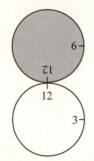

right clock has rotated 180°
(it's upside down), yet it has
made only 1/4 of its journey
around the left clock

FIGURE 53

55. a. Assume the babies are named A, B, C, and D. If two babies are correctly tagged, the other two labels must be switched for their labels to be incorrect. Thus the problem boils down to counting the number of ways two of the babies can be correctly tagged. Listing all the options is the easiest way to count them. Babies AB could be correctly labeled (leaving babies C and D labeled incorrectly), and the other options for correct labels are AC, AD, BC, BD, and CD. This makes six different ways.

b. If three of the babies have correct labels, the other baby must also have a correct label, which means there is no way to label three of the babies correctly while the fourth is labeled incorrectly.

UNIT 3A

QUICK QUIZ

1. **b**. A quantity triples when it is increased by 200%.

3. **c**. A negative relative change corresponds to a decreasing quantity.

5. **c**. 120% of $10 is $12.

7. **a**. The relative change in interest rates that go from 6% to 9% is $\frac{9\% - 6\%}{6\%} = 0.5 = 50\%$.

9. **c**. 10% of $1000 is $100, so your monthly earnings after one year will be $1100. Each year thereafter, your earnings will increase by more than $100 per year (because 10% of $1100 is more than $100), and the end result will be more than $1500.

DOES IT MAKE SENSE?

17. Makes sense. If a country's population is declining, then the percent change will be negative (and several of Europe's countries have, indeed, experienced recent declines in population).

19. Makes sense. An older child could easily weigh 25% more than a younger child (think of 100-pound and 125-pound children).

21. Does not make sense. Consider the case where I earn $100, and you earn $120. You earn $20 more than I do, which corresponds to 20% more than me (because 20% of $100 is $20). Even though I earn $20 less than you, I do not earn 20% less than you (because 20% of $120 is not $20). In fact, I earn 1/6 less than you do, which is about 16.7% less.

23. Makes sense. This just means that the incidence of cancer for children living near toxic landfills was seven times larger than the incidence of cancer among other children.

25. Makes sense. If the rate of return on the fund used to be 10%, then a 50% increase would result in a rate of return of 15%.

BASIC SKILLS AND CONCEPTS

27. $2/5 = 0.4 = 40\%$.

29. $0.20 = 1/5 = 20\%$.

31. $150\% = 1.5 = 3/2$.

33. $4/9 = 0.444\ldots = 44.44\ldots\%$.

35. $5/8 = 0.625 = 62.5\%$.

37. $69\% = 0.69 = 69/100$.

39. $7/5 = 1.4 = 140\%$.

41. $4/3 = 1.333\ldots = 133.33\ldots\%$.

43. 9 to 3 = 3. 3 to 9 = 1/3. 9 is 300% of 3. Note that the ratio of B to A is just the reciprocal of the ratio

of A to B, and that 300% is simply the ratio of A to B expressed as a percent.

45. 16 to 48 = 1/3. 48 to 16 = 3. 16 is 33.33…% of 48.

47. 440 to 8 = 55. 8 to 440 = 1/55. 440 is 5500% of 8.

49. 8.1 million to 21 million = 0.386. 21 million to 8.1 million = 2.593. 8.1 million is 38.6% of 21 million.

51. 4.1 million to 2.4 million = 1.71. 2.4 million to 4.1 million = 0.59. 4.1 million is 171% of 2.4 million.

53. 1500 square feet to 2300 square feet = 0.65. 2300 square feet to 1500 square feet = 1.53. 1500 square feet is 65% of 2300 square feet.

55. 220 is 51.2 % of 430 because 220/430 = 0.512.

57. 145 is 33.3% of 436 because 145/436 = 0.333.

59. 2.8 million is 16.8% of 16.7 million because 2.8/16.7 = 0.168.

61. John's absolute change in salary was $28,000 − $20,000 = $8000, and the relative change was $\frac{\$28{,}000 - \$20{,}000}{\$20{,}000} = 0.4 = 40\%$. Mary's absolute change was $35,000 − $25,000 = $10,000, and the relative change was $\frac{\$35{,}000 - \$25{,}000}{\$25{,}000} = 0.4 = 40\%$.

Thus Mary's salary grew more in absolute terms, but the relative change was the same. This happened because we are comparing Mary's larger absolute change to her also larger initial salary.

63. The absolute change is 300 million − 76 million = 224 million people. The relative change is $\frac{300{,}000{,}000 - 76{,}000{,}000}{76{,}000{,}000} = 2.95 = 295\%$.

65. The absolute change is 1450 papers − 2226 papers = −776 papers. The relative change is $\frac{1450 - 2226}{2226} = -0.349 = -34.9\%$.

67. The absolute difference is 2.20 million papers − 1.12 million papers = 1.08 million papers. The relative difference is

$$\frac{2{,}200{,}000 - 1{,}120{,}000}{1{,}120{,}000} = 0.964 = 96.4\%,$$

which means *USA Today*'s circulation is 96.4% more than the circulation of the *New York Times*.

69. The absolute difference is 77 million arrivals − 50 million arrivals = 27 million arrivals. The relative difference is $\frac{77 - 50}{50} = 0.54 = 54\%$, which means there were 54% more arrivals in France than in Spain.

71. The relative difference is 593,700 injuries − 535,100 injuries = 58,600 injuries. The relative difference is $\frac{593{,}700 - 535{,}100}{535{,}100} = 0.11 = 11\%$, which means there were 11% more basketball related injuries than bicycle related injuries.

73. Will's height is 122% of Wanda's height because 22% *more than* expresses a relative difference in heights, and this is equivalent to 100% + 22% *of* the reference height (i.e. Wanda's height).

75. Minnesota's population is 91% of Maryland's population because 9% *less than* expresses a relative difference in population, and this is equivalent to 100% – 9% *of* the reference population (i.e. Maryland's population).

77. Since the wholesale price is 40% *less than* the retail price, it is 60% *of* the retail price, which means it is 0.6 times the retail price.

79. Since the retail cost is 30% *more than* the wholesale cost, it is 130% *of* the wholesale cost, which means it is 1.3 times the wholesale cost.

81. The absolute change is 13.2% – 9.2% = 4 percentage points. The relative change is $\frac{13.2-9.2}{9.2}=0.435=43.5\%$.

83. The absolute change is 60% – 39% = 21 percentage points. The relative change is $\frac{60-39}{39}=0.538=53.8\%$.

85. In the first case, 33% of city employees in Freetown ride the bus to work because 10% *more than* 30% is 33%. In the second case, 40% of the city employees in Freetown ride the bus, because 40% is 10 percentage points more than 30%.

87. The final cost is $760 + 0.076 \times $760 = $817.76.

89. The final cost is $19,200 \times 1.072 = $20,582.40.

91. The cost for the dinner before the tip was $76.40/1.20 = $63.67.

93. The glass holds 4.17 ounces of wine, because 12% of 4.17 is 0.5 ounce. One can arrive at this answer by noting that 12% of the volume (call it V) is 0.5 ounce, which translates to $0.12V = 0.5$. Divide 0.5 by 0.12 to find V.

95. False. If Sue's weight began at W pounds, and she lost 15% of that weight, she would weigh $0.85W$ pounds. If she now gains 15%, her new weight would be $1.15 \times 0.85W = 0.9775W$ pounds. This is 2.25% less than her original weight.

97. False. Assign an arbitrary number to your property taxes before the increases, such as $100. If taxes increase by 6% in the first year, you'll pay $106. If they increase again by 7% in the next year, you'll pay $1.07 \times $106 = $113.42. This is an increase of 13.42%, not 13%.

99. Not possible. A price can decrease at most by 100%. Once you've deducted 100% from a price, it's free.

101. Possible. An increase of 400% means the price of real estate is five times what it used to be.

103. Possible. This means your computer is five times faster than my computer.

105. No, the mean score for both classes is not 85%. Assuming a 100-point exam, the first class scored a total of $25 \times 86 = 2150$ points, and the second class scored a total of $30 \times 84 = 2520$ points. Thus the combined number of points scored by 55 students is $2150 + 2520 = 4670$, and the mean score for both classes is $4670/55 = 84.9\%$.

FURTHER APPLICATIONS

107. True. 10% of 60% is 1/10 of 60%, which is 6%.

109. False. Some of the hotels may have both a pool and a restaurant, and those would be counted twice if the percentages were just added together.

111. An increase of 466% means Internet use in 2004 was 566% *of* the level in 2001 (call it x). This translates into $5.66x = 7.23\%$. Divide by 5.66 to find x: $x = 7.23\%/5.66 = 1.28\%$.

113. Add 20% of $3 billion to $3 billion to get $3.6 billion ($3 \times 1.2 = 3.6$).

115. The percent of people worried last winter (call it W) is up 16%, which means 116% *of* W is the new level of 80%. This translates to $1.16W = 80\%$. Divide by 1.16 to find W. $W = 80\%/1.16 = 69\%$.

117. Within three years, 60% *of* the 625,000 released prisoners will have been rearrested, which amounts to $0.60 \times 625,000 = 375,000$ rearrested.

119. The price of Oracle's stock in the previous week (call it P) increased by 3%, which means 103% *of* P is equal to $13.48. This translates to $1.03P = $13.48, and thus $P = $13.09 (divide by 1.03).

UNIT 3B

QUICK QUIZ

1. **a.** The decimal point must move seven places to the left to transform 70,000,000 into scientific notation, and thus its value is 7×10^7.

3. **c.** 10^9 is larger than 10^5 by a factor of 10^4, which is ten thousand.

5. **c.** A dollar bill is about 6 inches long, or about 15 cm long, and thus 8 million of them amounts to

$$8 \times 10^6 \text{ bills} \times \frac{1.5 \times 10^1 \text{ cm}}{\text{bill}} \times \frac{1 \text{ km}}{10^5 \text{ cm}} = 12 \times 10^2 \text{ km} ,$$

which is 1200 kilometers. This is enough to get into outer space, but not near enough to reach the moon.

7. **b.** If 1 inch = 20 miles, then 6 inches is $6 \times 20 = 120$ miles.

9. **c.** Assuming you're willing to campaign for eight hours per day (which is roughly 500 minutes), and that you'll limit yourself to 10 minutes of talking at each house (including travel time between houses), you'll be able to visit 50 households per

day, or 100 houses in two days. This translates into 500 houses in two weeks (campaigning five days per week), or 1000 households in a month, or 12,000 households in a year-long campaign, where you do nothing but campaigning. Clearly it would be impossible to carry out your plan before the election – pay for some TV advertisements instead.

DOES IT MAKE SENSE?

19. Makes sense. The page you are reading right now has two columns, each of which has about 50 lines, and there are roughly 10 words per line, for a total of 1000 words per page. A book with $10^5 = 100,000$ words would have 100 pages. A paperback with pages about half the size of this solutions manual would be two-hundred pages long.

21. Makes sense. At ten feet per floor, an apartment building with 20 floors would be 200 feet tall.

23. Does not make sense. A typical NFL stadium holds 70,000 fans. At an unrealistic speed of 10 signatures per minute, it would take the football star 7000 minutes (roughly 100 hours) to provide signatures for all the fans (who would grow restless waiting in line for a signature after about 15 minutes).

BASIC SKILLS AND CONCEPTS

25. a. $3 \times 10^3 = 3000 =$ three thousand .

 b. $6 \times 10^6 = 6,000,000 =$ six million .

 c. $3.4 \times 10^5 = 340,000 =$ three hundred forty thousand.

 d. $2 \times 10^{-2} = 0.02 =$ two hundredths .

 e. $2.1 \times 10^{-4} = 0.00021 =$ twenty-one hundred-thousandths.

 f. $4 \times 10^{-5} = 0.00004 =$ four hundred - thousandths.

27. a. $233 = 2.33 \times 10^2$.

 b. $126,547 = 1.26547 \times 10^5$.

 c. $0.11 = 1.1 \times 10^{-1}$.

 d. $9736.23 = 9.73623 \times 10^3$.

 e. $124.58 = 1.2458 \times 10^2$.

 f. $0.8642 = 8.642 \times 10^{-1}$.

29. a. $(3 \times 10^3) \times (2 \times 10^2) = 6 \times 10^5$.

 b. $(4 \times 10^2) \times (3 \times 10^8) = 12 \times 10^{10} = 1.2 \times 10^{11}$.

 c. $(3 \times 10^3) + (2 \times 10^2) = 3000 + 200 = 3200$

 $= 3.2 \times 10^3$.

 d. $(8 \times 10^{12}) \div (4 \times 10^4) = 2 \times 10^8$.

31. a. 10^{35} is 10^9 (1 billion) times as large as 10^{26} .

 b. 10^{27} is 10^{10} (10 billion) times as large as 10^{17} .

 c. 1 billion is 10^3 (1000) times as large as 1 million.

 d. 7 trillion is 10^9 (1 billion) times as large as 7 thousand.

 e. 2×10^{-6} is 10^3 (1000) times as large as 2×10^{-9} .

 f. 6.1×10^{29} is 10^2 (100) times as large as 6.1×10^{27} .

33. Total energy consumption in the United States is about 1×10^{20} joules.

35. The hard drive on my computer has a capacity of 1.2×10^{10} bytes.

37. The diameter of a typical bacterium is about 1×10^{-6} meter.

39. a. 300,000 is 3×10^5 , and 100 is 1×10^2 , so their product is 3×10^7 , or 30,000,000. This is an exact answer as no estimation was necessary.

 b. 5.1 million is about 5×10^6 , and 1.9 thousand is about 2×10^3 , so their product is about 1×10^{10} , which is 10 billion. The exact answer (computed with a calculator) is 9.69 billion.

 c. Since 2.1×10^6 is about 2 million, $4 \times 10^9 \div 2.1 \times 10^6$ is about 2×10^3 , or 2000. The exact answer is nearer to 1.9×10^3 (1904.76 is a rounded answer).

 d. 33 multiplied by 3.1 is very close to 100, so $(33 \times 10^6) \times (3.1 \times 10^3)$ is about 100×10^9 , which is 1×10^{11} . The exact answer is 1.023×10^{11} .

 e. 4,288,364 is about 4×10^6 , and 2132 is about 2×10^3 , so their quotient is about 2×10^3 , or 2000. The exact answer is 2011 (rounded).

 f. $(6.129845 \times 10^6) \div (2.198 \times 10^4)$ is about $(6 \times 10^6) \div (2 \times 10^4) = 3 \times 10^2$, or 300. The exact answer is 279 (rounded).

41. Answers will vary regarding the estimates made, but most people spend more money on gas each year than on movies. If you see two movies per month, spending $10 per movie, that's $20 per month, which amounts to $240 per year. On the other hand, if you fill up your gas tank twice per month, spending $30 to fill a tank, that's $60 per month, which amounts to $720 per year.

43. Assuming you sleep 7 hours per day, and that you study less than 7 hours per day, you'll sleep more than you study. In particular, if you sleep 7 hours each night, you'll sleep 2555 hours over the course of a year. If you study 6 hours per day over the course of a school year (30 weeks, or two semesters), you'll study 1260 hours per year.

45. Answers will vary, but if one-third of the population (100 million people) each spend $100

per year on movies, that would produce $10 billion in revenue.

47. Answers will vary, but assuming a daily water intake of 2 pints, one can estimate the annual water consumption with a unit conversion:

$1 \text{ yr} \times \dfrac{365 \text{ d}}{\text{yr}} \times \dfrac{2 \text{ pt}}{\text{d}} \times \dfrac{1 \text{ qt}}{2 \text{ pt}} \times \dfrac{1 \text{ gal}}{4 \text{ qt}}$, which is about 90 gallons per year.

49. Insurance companies use the figure of 15,000 miles per year for the average distance driven by an adult. If your car gets 25 miles to the gallon, you'll use about 600 gallons each year. Answers will vary, of course.

51. Table 3.1 states that 4 million joules are required for one hour of running, and that an average candy bar supplies 1 million joules of energy. Thus four candy bars would be needed for each hour of running, which means you'd need to eat 24 candy bars to supply the energy for six hours of running.

53. One kilogram of uranium-235 releases 5.6×10^{13} joules, which is 35,000 times as much as the energy released by one kilogram of coal (1.6×10^{9} joules): $\dfrac{5.6 \times 10^{13}}{1.6 \times 10^{9}} = 35,000$.

55. $\dfrac{5 \times 10^{7} \text{ joules}}{\text{home}} \times \dfrac{1 \text{ liter } H_2O}{7 \times 10^{13} \text{ joules}} = 7 \times 10^{-7} \dfrac{\text{liters}}{\text{home}}$. This is just a little less than one millionth of a liter.

57. $\dfrac{1 \times 10^{20} \text{ joules}}{\text{yr}} \times \dfrac{1 \text{ kg uranium}}{5.6 \times 10^{13} \text{ joules}} = 1.8 \times 10^{6} \dfrac{\text{kg}}{\text{yr}}$. Thus 1.8 million kg of uranium would be needed to supply the annual energy needs of the United States.

59. Since there are 100,000 cm in 1 km, the scale is 100,000 to 1.

61. Since 2 inches represents 2 miles, 1 inch represents one mile, and as there are 63,360 inches in one mile ($1 \text{ mi} \times \dfrac{5280 \text{ ft}}{1 \text{ mi}} \times \dfrac{12 \text{ in.}}{1 \text{ ft}} = 63,360 \text{ in.}$), the scale is 63,360 to 1.

63. Since there are 20,000,000 cm in 200 km, the scale is 20,000,000 to 5, or 4,000,000 to 1.

65. There are about 820 ft in 250 meters (see below), so the scale is 820 to 1.

$250 \text{ m} \times \dfrac{100 \text{ cm}}{\text{m}} \times \dfrac{1 \text{ ft}}{30.48 \text{ cm}} = 820 \text{ ft}$.

67. In order to scale down each of the actual diameters and distances listed in the table, begin by dividing by 10,000,000,000 (10 billion). Since this produces tiny answers when units of kilometers are used, it is best to convert to a more convenient unit. The following table shows diameters listed in millimeters (rounded to the nearest tenth of a millimeter), and distances listed in meters (rounded to the nearest meter).

Planet	Model Diameter	Model distance from Sun
Mercury	0.5 mm	6 m
Venus	1.2 mm	11 m
Earth	1.3 mm	15 m
Mars	0.7 mm	23 m
Jupiter	14.3 mm	78 m
Saturn	12.0 mm	143 m
Uranus	5.2 mm	287 m
Neptune	4.8 mm	450 m

69. The scale on this timeline would be 14 billion years to 100 meters, and thus 1 billion years is equivalent to 1/14 of 100 meters, which amounts to 7.1 meters. Since 10,000 years is 1/100,000 of one billion years, we only need to divide 7.1 meters by 100,000 to find the distance that goes with 10,000 years. This gives 0.000071 meters, or 0.071 millimeters.

FURTHER APPLICATIONS

71. $\dfrac{1.4 \times 10^{6} \text{ crimes}}{\text{yr}} \times \dfrac{1 \text{ yr}}{365 \text{ d}} \times \dfrac{1 \text{ d}}{24 \text{ hr}} = 160 \dfrac{\text{crimes}}{\text{hr}}$.

73. There are roughly 300 million Americans, and since 300 million metric tons of corn are produced, there's one metric ton per person.

75. The thickness of a stack of ten $1 bills is about one millimeter, and this can be used to convert $8.5 trillion into a height.

$\$8.5 \times 10^{12} \times \dfrac{1 \text{ mm}}{\$10} \times \dfrac{1 \text{ km}}{10^{6} \text{ mm}} = 8.5 \times 10^{5} \text{ km}$.

77. $1.35 \times 10^{6} \text{ students} \times \dfrac{1 \text{ stadium}}{70,000 \text{ students}} = 19 \text{ stadiums}$.

79. Answers will vary, but if you assume there are 200 million drivers in the U.S. (2/3 of the population), and each uses 1000 gallons of gas per year, then roughly 2 trillion kg of carbon dioxide was added to the atmosphere over the past year:

$2 \times 10^{8} \text{ drivers} \times \dfrac{1000 \text{ gal}}{\text{driver}} \times \dfrac{10 \text{ kg}}{\text{gal}} = 2 \times 10^{12} \text{ kg}$.

81. a. Energy = power × time, so we need to multiply the solar power by one second, and then take 0.06% of the result, followed by 1% of that result to account for the fact that not all solar power is converted into plant matter.

$180,000 \times 10^{12} \text{ W} = 1.8 \times 10^{17} \dfrac{j}{s} \times 1 \text{ s} = 1.8 \times 10^{17} \text{ j}$. This is the amount of energy that reaches the earth each second. Multiply by 0.0006, and then by 0.01 (the decimal equivalent of 0.06% and 1%) to find the amount of energy stored in plant matter each second – this results in 1.08×10^{12} joules.

b. If 1.08×10^{12} joules of energy are used every second, that's the same as 1.08×10^{12} watts, or about 1 terawatt. This is only 10% of the world's electricity demand, so it would not be enough.

c. To avoid depleting plant life, it's necessary to use fossil fuels, which provide the energy stored by plants over thousands of years.

83. Answers will vary, but if a ream of paper (500 sheets) weighs 8 pounds, then each sheet will weigh 1/500 of 8 pounds, which is 0.016 lb.

85. Answers will vary considerably, depending upon the sand that is used to make the estimate.

UNIT 3C

QUICK QUIZ

1. **b**. The best economists can do when making projections into the future is to base their estimates on current trends, but when these trends change, it often proves the predictions wrong.

3. **b**. 1.020 has four significant digits, 1.02 has three, and 0.000020 has only two.

5. **b**. If you place the thermometers in sunlight, this is a problem with your system of measurement, and the readings will all be too high, which is a systematic error.

7. **b**. The absolute error in all cases was -50 points, but unless all students had the same score (very unlikely), the relative errors will be different.

9. **a**. Reporting the debt to the nearest penny is very precise, though the debt is so large that we really don't know it to that level of precision, and thus it is not likely to be very accurate.

DOES IT MAKE SENSE?

17. Does not make sense. Predicting the federal deficit for the next year to the nearest million dollars is nearly impossible.

19. Does not make sense. Unless you have access to an exceptionally precise scale, you could never measure your weight to that level of precision (and even with an excellent scale, your weight changes at the level reported when you sweat, breath, lose a few hairs, etc.).

21. Does not make sense. A typical yard stick may be precise to the nearest 1/16 of an inch, but certainly not to the nearest micrometer.

23. Makes sense. Provided the company's revenue is $1 billion, $1 million would be only 0.1% of revenue, and it wouldn't be considered a large error.

BASIC SKILLS & CONCEPTS

25. a. 3.   b. 88.   c. 0.
    d. 185.   e. 1945.   f. 3.
    g. 6.   h. 1500.   i. −14.

27. There are four significant digits, and it is precise to the nearest dollar.

29. There is one significant digit, and it is precise to the nearest 10 mph.

31. There are six significant digits, and it is precise to the nearest 0.00001 mile.

33. There are two significant digits, and it is precise to the nearest 1000 seconds.

35. There are seven significant digits, and it is precise to the nearest 0.0001 pound.

37. There are four significant digits, and it is precise to the nearest 0.1 km/s.

39. $45 \times 32.1 = 1440$.

41. $231.89 \div 0.034 = 6800$.

43. $(2.3 \times 10^5) \times (7.963 \times 10^3) = 1.83149 \times 10^3$.

45. Random errors may be introduced when the cars are not counted properly, and systematic errors could occur due to interpretation of what constitutes a sport utility vehicle.

47. Random errors could occur if the average is calculated incorrectly. Systematic errors might occur if people repeatedly understate their incomes to avoid higher taxes.

49. Random errors could occur by misreading the scale. A minor systematic error could occur if deli employees routinely measured the weight with the meat already in the deli bag.

51. Random errors may occur due to miscounting of kernels, and a systematic error could occur if the clerk overfills the box for friends.

53. Problem (1) is a random error because the mistakes may be too high or two low. Problem (2) is a systematic error, because the tax payer likely underreported his income.

55. This is a systematic error that results in all altitude readings being 2780 lower than they should be.

57. The absolute error is measured value – true value, or $19.00 – $18.50 = $0.50. The relative error is
$$\frac{\text{measured value} - \text{true value}}{\text{true value}} = \frac{\$19 - \$18.50}{\$18.50} = 0.027 = 2.7\%.$$

59. The absolute error is $39.20 – $43.24 = –$4.04. The relative error is ($39.20 – $43.24)/$43.24 = –0.093 = –9.3%.

61. Regarding 6 cups as the true value (because you measured 6.25 cups), the absolute error is 6.25 cups – 6 cups = 0.25 cup. The relative error is (6.25 cups – 6 cups)/6 cups = 0.042 = 4.2%.

63. The absolute error is 9 hr – 10 hr = –1 hr. The relative error is –1 hr/10 hr = –0.1 = –10%.

65. The measurement obtained by the tape measure is more accurate because the value is closer to your true height than the value obtained at the doctor's office (1/8 inch versus 0.4 inch). However, the laser at the doctor's office is more precise as it measures to the nearest 0.05 (or 1/20) inch, while the tape measure measures to the nearest 1/8 inch.

67. The digital scale at the gym is both more precise and more accurate. It is more precise because it measures to the nearest 0.01 kg, while the health clinic's scale measures to the nearest 0.5 kg, and it is more accurate because its reported weight is nearer to your true weight than the scale at the health clinic.

69. A calculator gives 60 hr – 1.45 hr = 58.55 hr. This should be rounded to the nearest 10 hr because that's the level of precision of 60 hr (we can't be sure whether the 0 in 60 is a significant digit, so we must assume that it is not). The final answer is 60 hr.

71. A calculator gives 62.5 km/hr × 2.4 hr = 150 km. The second measurement has two significant digits, and so does our final answer, so it needn't be changed.

73. The least precise measurement is 36 miles, which is precise to the nearest mile, so the answer should be rounded to the nearest mile. 36 mi + 2.2 mi = 38.2 mi, which is 38 mi when rounded to the correct level of precision.

75. The cost is $2.1 million/120,345 people = $17.45 per person. However, as the first measurement has only two significant digits, so must the answer, so we round to $17 per person.

FURTHER APPLICATIONS

77. This measurement could have many sources of errors, both random and systematic (for example, transient people might get counted twice, and people on the frontier – remember, this is 1820 – might be undercounted). The measurement is not believable as the level of precision (to the nearest person) is unattainable. In the time it would take to conduct the survey, many people would have been born and died, and people immigrating to the country change its population on a daily basis.

79. Systematic errors could arise depending on how the length of a river is defined and how it is measured (should we measure along the banks, or right down the middle?), and random errors could easily arise due to changes in the course of the river over time, or errors in recording the data. The fact is not believable because (as explained above), it is rather difficult to define how the length of a river should be measured, let alone measure such a long river to the nearest mile.

81. Random and systematic errors could be present in the temperature readings that lead to the average high of 88 degrees. The fact is believable to the level of precision given.

83. The mathematical models used to estimate the mass of the earth could introduce systematic errors, and random errors could be introduced in the experiments conducted to find the mass. The mass is believable to the level of precision given.

UNIT 3D

QUICK QUIZ
1. **b.** The 1985 index of 210.9 is a little more than twice as large as the 1975 index of 100, so the price of gas in 1985 is roughly twice as much as the price in 1975.
3. **b.** Like most indexes, the CPI is designed to allow one to compare prices from one year to another.
5. **c.** The CPI in 2005 was 195.3, and it was 152.4 in 1995. Thus prices in 2005 are 28% more than prices in 1995 (because 195.3/152.4 = 1.28). If the CPI were recalibrated so that 1995 = 100, the new CPI in 2005 would be 128, which is lower than 195.3.
7. **a.** If the cost of college has increased faster than the CPI, but one's salary has only kept pace with the CPI, then the cost of college as a percentage of salary has increased, making it more difficult to afford.
9. **c.** Measured in 2006 dollars, gas was less than $1.50 per gallon in 1998–1999, and this is lower than any other time shown on the chart.

DOES IT MAKE SENSE?
15. Does not make sense. One must consider the affect of inflation on the price of gas, and compare the prices in, say, 2006 dollars in order to make a valid comparison.
17. Makes sense. A penny in Franklin's day would be worth something on the order of a dollar today.
19. Does not make sense. The reference year that is chosen for an index number does not affect the trends of inflation, so if milk is more expensive in 1995 dollars, it should also be more expensive in 1975 dollars.

BASIC SKILLS & CONCEPTS
21. To find the index number, take the current price and divide it by the 1975 price: $2.80/$0.567 = 4.938 = 493.8%. The price index number is 493.8 (the percentage sign is dropped).

23. The price index number in 1995 is 212.5, which means prices in 1995 were 212.5% of the prices in 1975. Thus $10 spent on gas in 1975 is equivalent to $10 × 212.5% = $10 × 2.125 = $21.25.

25. With 1965 as the reference value, the price index number for that year is 100. To find the index for 1955, divide the 1955 price by the 1965 price, convert to a percent, and drop the percent sign:

$$\frac{29.1¢}{31.2¢} = 0.933 = 93.3\%,$$ so the index is 93.3.

Follow the same procedure to find the index for 1975: $$\frac{56.7¢}{31.2¢} = 1.817 = 181.7\%,$$ so the index is 181.7. Do the same thing for the remaining years to arrive at the following index numbers. 1985 = 383.3. 1995 = 386.2. 2000 = 496.8. 2005 = 740.4.

27. Divide the 2005 CPI by the 1975 CPI to find the factor by which prices in 2005 were larger than prices in 1975: $$\frac{CPI_{2005}}{CPI_{1975}} = \frac{195.3}{53.8} = 3.63.$$ This means prices in 2005 were 3.63 times larger than prices in 1975, so $20,000 in 1975 is equivalent to $20,000 × 3.63 = $72,600 in 2005.

29. Inflation is the relative change in the CPI, which is $$\frac{CPI_{1978} - CPI_{1977}}{CPI_{1977}} = \frac{65.2 - 60.6}{60.6} = 0.076 = 7.6\%.$$

31. Because $$\frac{CPI_{2004}}{CPI_{1979}} = \frac{188.9}{72.6} = 2.6,$$ prices in 2004 were 2.6 times as large as prices in 1979. Thus a 75¢ loaf of bread would be $0.75 × 2.6 = $1.95 in 2004 dollars.

33. Because $$\frac{CPI_{1975}}{CPI_{2000}} = \frac{53.8}{172.2} = 0.3124,$$ prices in 1975 were 0.3124 times those in 2000. Thus a $7.00 movie would be $7.00 × 0.3124 = $2.19 in 1975 dollars.

35. The purchasing power of a 1973 dollar in terms of 2002 dollars is $$\$1 \times \frac{CPI_{2002}}{CPI_{1973}} = \$1 \times \frac{179.9}{44.4} = \$4.05.$$

37. In Palo Alto, the price would be $$\$300,000 \times \frac{365}{87},$$ which is $1.26 million. In Sioux City, the price would be $$\$300,000 \times \frac{47}{87} = \$162,000.$$ In Boston, the price would be $$\$300,000 \times \frac{182}{87} = \$627,600.$$

39. In Spokane, the price would be $$\$250,000 \times \frac{78}{75},$$ which is $260,000. In Denver, the price would be $$\$250,000 \times \frac{87}{75} = \$290,000.$$ In Juneau, the price would be $$\$250,000 \times \frac{100}{75} = \$333,000.$$

41. Health care spending increased by a percent of $$\frac{\$1.8 \times 10^{12} - \$80 \times 10^9}{\$80 \times 10^9} = 21.5 = 2150\%,$$ whereas the overall rate of inflation was $$\frac{CPI_{2004} - CPI_{1973}}{CPI_{1973}} = \frac{188.9 - 44.4}{44.4} = 3.25 = 325\%.$$

43. The relative change in the cost of college was $$\frac{\$27,516 - \$5900}{\$5900} = 3.66 = 366\%.$$ The rate of inflation was $$\frac{CPI_{2004} - CPI_{1980}}{CPI_{1980}} = \frac{188.9 - 82.4}{82.4} = 1.29,$$ which is an increase of 129%.

45. The relative change in the price of a home was $$\frac{\$155,500 - \$75,300}{\$75,300} = 1.07 = 107\%.$$ The rate of inflation was $$\frac{CPI_{2004} - CPI_{1990}}{CPI_{1990}} = \frac{188.9 - 130.7}{130.7} =,$$ 0.45 which is an increase of 45%.

47. As shown in the first line of the table, $0.25 in 1938 dollars is worth $2.78 in 1996 dollars.

49. In 1996, actual dollars are 1996 dollars.

51. The 1979 minimum wage, measured in 1979 dollars, was $2.90. To convert this to 1996 dollars, use the CPI from 1996 and 1979.

$$\$2.90 \times \frac{CPI_{1996}}{CPI_{1979}} = \$2.90 \times \frac{156.9}{72.6} = \$6.27.$$

This agrees with the entry in the table.

53. The purchasing power of the minimum wage was highest in 1968, when its value in 1996 dollars was highest ($7.21). Since all entries in the last column are listed in 1996 dollars, we can make valid comparisons between values in the column.

UNIT 3E

QUICK QUIZ

1. **a.** As stated in the text surrounding the table, the old treatment had a better overall treatment rate, even though the new treatment performed better in each category.

3. **b.** Even though Derek had a higher GPA in each of the two years, we can't be sure his overall GPA was higher than Terry's – Simpson's paradox may be at play.

5. **c.** A *false positive* means the test detected steroids (that's the *positive* part), but the person never used steroids (that's the *false* part).

7. **b.** Those who had false negatives (15) and false positives (1485) make up the group who had incorrect test results.

9. **b**. A person earning $200,000 will save 2.9% of his income, which is $5800. A person earning $50,000 will save 4.4% of his income, which is $2200. Thus the person earning $200,000 or more will save more total tax dollars than someone earning between $40,000 and $50,000.

DOES IT MAKE SENSE?

15. Makes sense. If Simpson's paradox is at work, both categories could show an improvement with the new drug even though overall, the old drug may have done a better job.

17. Does not make sense. There are scenarios (see discussion in text concerning mammograms and drug tests, for example) where a positive test may not correlate well with the probability that a bag contains banned materials, even though the test is 98% accurate.

19. Does not make sense. Both sides can make a valid argument for their position (see discussion in text about tax cuts).

BASIC SKILLS & CONCEPTS

21. As shown in the tables, Josh had a higher batting average in both the first and second half of the season. Jude had the higher overall batting average (80/200 = .400 versus 85/220 = .386). This illustrates Simpson's paradox because it's a case where one person (Josh) performed better in two of two categories (the first and second halves of the season), and yet Jude outperformed Josh over the course of the entire season.

23. a. New Jersey had higher scores in both categories (283 versus 281, and 252 versus 250), but Nebraska had the higher overall average (277 versus 272).
b. The explanation lies in Simpson's paradox: it's possible for one state (Nebraska) to score lower in both categories, and yet have a higher overall percentage. This is due to the different racial makeup of the two state's populations. The white students are scoring better than the nonwhite students in both states, and Nebraska has a larger percentage of white students than New Jersey, so they influence the overall scores more heavily than the white students in New Jersey.
c. The verification is an exercise in a weighted average. 87% of the population had an average of 281, and 13% of the population had an average of 250, so the overall average is 0.87(281) + 0.13(250) = 277.
d. As in part **c**, we only have to compute a weighted average: 0.66(283) + 0.34(252) = 272.
e. See part **b**.

25. a. The death rate for whites was 8400/4,675,000 = 0.0018 = 0.18%. For nonwhites, it was 500/92,000 = 0.0054 = 0.54%. Overall, the death rate was 8900/4,767,000 = 0.0019 = 0.19%.
b. The death rate for whites was 130/81,000 = 0.0016 = 0.16%. For nonwhites, it was 160/47,000 = 0.0034 = 0.34%. Overall, the death rate was 290/128,000 = 0.0023 = 0.23%.
c. Simpson's paradox is at work here because the death rates for TB in each category (white and nonwhite) were higher in New York City than in Richmond, and yet the overall death rate due to TB was higher in Richmond. The paradox arises because Richmond had a different racial makeup than New York City: it had a much higher percentage of nonwhite residents than New York, and as the death rate for TB was much higher among nonwhites than whites in both cities, the higher percentage of nonwhites in Richmond has a more pronounced influence on the death rate.

27. a. There are 10,000 total participants with tumors in the study (lower right hand entry), and 1% of them (100) have malignant tumors. Since the test is 90% accurate, 90 of those with malignant tumors will be true positives, while the other 10 will be false negatives. 99% of the women (9900) will have benign tumors. 90% of these (8910) will be true negatives, while 10% of these (990) will be false positives. The only numbers in the table not yet verified are the first two entries in the Total column; these are simply the sum of the rows in which they lie.
b. Of the 1080 who have positive mammograms, only 90 are true positives, so only 90/1080 = 8.3% of them actually have cancer.
c. 100 of the women have cancer (malignant tumors), and 90 of those will have a true positive test result, which is 90%. Thus if you really have cancer, the test will detect it 90% of the time.
d. Of the 8920 who have a negative mammogram, only 10 of them are false negatives, which translates to a probability of 10/8920 = 0.11%.

29. a. Beginning with the last row, the sample size is 4000, and 1.5% of them (60) have the disease, while the remaining 3940 do not. The first column shows the 60 with the disease, and the test detects 90% of them (54), but misses 6 of them. The second column shows the 3940 who do not have the disease, and the test says 90% of them (3546) do not have the disease (a negative test), but it says 10% of them (394) do have the disease. The entries in the last column are simply the sum of the rows in which they lie.
b. 54 of the 60 (90%) who have the disease test positive.

c. 54 of the 448 (12%) who test positive actually have the disease. In part **b**, the answer of 90% is the ratio of those who test positive out of those who have the disease, while the answer of 12% here is the ratio of those who have the disease among those that test positive.

d. As shown in part **c**, if you test positive, you have a 12% chance of actually having the disease. This is 8 times the overall incidence rate (1.5%), which means there is reason for concern, and yet it's still a fairly low chance that you actually have the disease.

FURTHER APPLICATIONS

31.   a. Spelman won 10/29 = 34% of its home games and 12/16 = 75% of its away games, while Morehouse won 9/28 = 32% of its home games and 56/76 = 74% of its away games. Since Spelman is better than Morehouse in each category, it's the better team.

b. Morehouse won 65/104 = 62.5% of all the games they played, while Spelman won 22/45 = 49% of all the games they played. Based on these figures, Morehouse is the better team.

c. Disregarding the strength of the opponents each college played, it's universally accepted that the team with a higher percentage of wins is the better team, and thus the claim in part **b** makes more sense.   Additionally, it's more difficult to win away games than home games, and while both teams have an impressive away-game win percentage, Morehouse complied its overall record of 62.5% with many more away games.

33.   a. Note that 10% of the 5000 at risk people is 500, and this is the sum of the first row in the first table. For the 20,000 people in the general population, note that 0.3% of 20,000 is 60, and this is the sum of the first row in the second table. 95% of the 500 infected people in the at risk group should test positive, and 5% should test negative – take 95% and 5% of 500 to verify the first row entries in the first table. 95% of the 4500 at risk people who aren't infected should test negative, while 5% should test positive – take 95% and 5% of 4500 to verify the second row in the first table. A similar process will show that the numbers in the second table are also correct.

b. 475/500 = 95% of those with HIV test positive. 475/700 = 67.9% of those who test positive have HIV. The percentages are different because the number of people in each group (those who have HIV and those who test positive) is different.

c. As shown in part **b**, if you test positive, and you are in the at risk group, you have a 67.9% chance of carrying the disease. This is much greater than the overall percentage of those who have the

disease (10%), which means if you test positive, you should be concerned.

d. 57/60 = 95% of those with HIV test positive. 57/1054 = 5.4% of those who test positive have HIV. The two percentages are different because the number of people in each group (those who have HIV and those who test positive) is different.

e. As shown in part **d**, if you test positive, and you are in the general population, you have a 5.4% chance of carrying the disease. This is larger than the overall percentage of those who have the disease (0.3%), but it's still small enough that it's best to wait for more tests before becoming alarmed.

35.   The data suggest a hiring preference for women because a higher percentage of women were hired in each of the two categories (white-collar and blue-collar jobs). On the other hand, if one looks at the overall picture, it turns out a higher percentage of male applicants was hired. (There were 20% of 200 applicants = 40 female workers hired for white-collar jobs, and 85% of 100 applicants = 85 female workers hired for blue-collar jobs. Thus there were 125 workers hired from a pool of 300 applicants, which is 125/300 = 41.7%. There were 15% of 200 applicants = 30 male workers hired for white-collar jobs, and 75% of 400 applicants = 300 male workers hired for blue-collar jobs. Thus there were 330 workers hired from a pool of 600 applicants, which is 330/600 = 55%.) This is an example of Simpson's paradox because females were hired at a greater rate in both categories of workers, and yet males were hired at a greater rate overall. The paradox can be resolved by realizing that there were many more males hired for the blue-collar jobs, which are more abundant than the white-collar jobs, and thus the male-dominated blue-collar jobs have a greater influence on the overall hiring rate.

41.   a. The absolute difference in savings is $12,838 – $211 = $12,627. A single person earning $41,000 will save $211/$41,000 = 0.5% of income. A single person earning $530,000 will save $12,838/$530,000 = 2.4% of income.

b. The absolute difference in savings is $13,442 – $1208 = $12,234. A married couple earning $41,000 will save $1208/$41,000 = 2.9% of income. A married couple earning $530,000 will save $13,442/$530,000 = 2.5% of income.

UNIT 4A

QUICK QUIZ

1. **a.** Evaluating your budget allows you to look critically at your cash flow, which affects personal spending.
3. **b.** If your cash flow is negative, you are spending more than you take in.
5. **b.** Housing typically costs about one-third of your income.
7. **a.** You can't save without money left over at the end of the month, which corresponds to a positive cash flow.
9. **c.** Kira spends $7.50 per day on soda, so over the course of a year, she spends nearly $3000.

DOES IT MAKE SENSE?

17. Does not make sense. All the smaller expenses do add up, and they have a significant influence on your budget.
19. Makes sense. When prorated as a per-month expense, an $1800 vacation costs $150 per month.
21. Does not make sense. Divide $15,000 by 365 (days per year) to find that an annual expense of $15,000 comes to $41 per day. Pizza and a soda do not cost that much.

BASIC SKILLS & CONCEPTS

23. Natasha spends $5 each week on lottery tickets, and using 52 weeks per year, this comes to $260 per year. She spends $120 each month on food, which is $1440 per year. The amount she spends on lottery tickets is $260/$1440 = 18% of the amount she spends on food.
25. Suzanne spends $85 per month for her cell phone usage, which is $1020 per year. This is $1020/$200 = 510% of the amount she spends on health insurance.
27. Sheryl spends $9 each week on cigarettes, which comes to $468 per year. She spends $30 each month on dry cleaning, or $360 per year. The amount she spends on cigarettes is $468/$360 = 130% of the amount she spends on dry cleaning.
29. Vern spends $21 per week on beer, which amounts to $1092 per year. This is $1092/$700 = 156% of the amount he spends on textbooks.
31. Since 18%/12 = 1.5%, you spend 1.5% of $650 on interest each month, which is $0.015 \times $650 = $9.75. Over the span of a year, you'll spend $117 on interest.
33. Vic's balance is $2200 − $300 = $1900. He spends 3% of $1900, or $0.03 \times $1900 = $57 each month on interest. This comes to $684 per year.

35. Sara spends $4800 per semester, or $9600 per year, which is $9600/12 = $800 per month.
37. At $280/credit, Moriah spends $280 \times 15 = $4200 per quarter on tuition. Add this to her quarterly fees ($190), and her dorm room charge ($2300), with the result that she spends $6690 per quarter. Over the span of a year, she spends $6690 \times 3 = $20,070, which means she spends $20,070/12 = $1673 per month.
39. Nguyen spends $200 + $400 \times 2 = $1000 each year on these expenses, or $1000/12 = $83 per month.
41. Income = $600 \times 12 + $400 \times 12 + $5000 = $17,000 per year. Expenses = $450 \times 12 + $50 \times 48 + $3000 \times 2 + $100 \times 48 = $18,600 per year. Thus annual cash flow is $17,000 − $18,600 = −$1600, and monthly cash flow is −$1600/12 = −$133.
43. Income = $2300 \times 12 = $27,600. Expenses = $800 \times 12 + $90 \times 48 + $125 \times 12 + $360 \times 2 + $400 \times 2 + $25 \times 48 + $400 \times 12 + $85 \times 12 = $23,960. Thus annual cash flow is $27,600 − $23,960 = $3640, and monthly cash flow is $3640/12 = $303.
45. The woman spends $900/$3200 = 28% of her income on rent (i.e. housing), which is below average.
47. The man spends $200/$3600 = 5.6% on health care, which is about average.
49. The couple spends $600/$4200 = 14% on health care, which is above average.
51. The cost of using the old car for five years is computed below.

Gas:
$$\frac{250 \text{ mi}}{\text{wk}} \times \frac{1 \text{ gal}}{21 \text{ mi}} \times \frac{\$3.50}{1 \text{ gal}} \times \frac{52 \text{ wk}}{\text{yr}} \times 5 \text{ yr} = \$10,833 \, .$$

Insurance: $\frac{\$400}{\text{yr}} \times 5 \text{ yr} = \$2000 \, .$

Repairs: $\frac{\$1500}{\text{yr}} \times 5 \text{ yr} = \$7500 \, .$

Total: $10,833 + $2000 + $7500 = $20,333.
The cost for using the new car for five years is computed below.

Gas: $\frac{250 \text{ mi}}{\text{wk}} \times \frac{1 \text{ gal}}{45 \text{ mi}} \times \frac{\$3.50}{1 \text{ gal}} \times \frac{52 \text{ wk}}{\text{yr}} \times 5 \text{ yr} = \$5056 \, .$

Insurance: $\frac{\$800}{\text{yr}} \times 5 \text{ yr} = \$4000 \, .$

Purchase Price: $16,000.
Total: $5056 + $4000 + $16,000 = $25,056.
Using the old car is less expensive, but not by a large factor. After five years, if you use the old car, you'll have a junky car, whereas if you bought the new car, you'd have a nice car after five years.
53. If you buy the car for $22,000, and sell it three years later for $10,000, you will have spent

$12,000. If you lease the car for three years (36 months) you will spend $1000 + $250 × 36 = $10,000. It is cheaper to lease in this case.

55. At the in-state college, you will spend $4000 + $700 × 12 = $12,400 each year. At the out-of-state college, you will spend $6500 + $450 × 12 = $11,900. It will cost less out-of-state.

57. If you take the course, you will spend $1500, and use 150 hours of time that could be used to earn money in a job. If you work those 150 hours, you will earn $1500.

FURTHER APPLICATIONS

59. Note that it costs $10 less per month to operate the new dryer, and thus in 62 months, you would have saved $620, which is long enough to pay for the new dryer.

61. a. If you didn't have the policy, you'd pay all of the costs of the claims, which is $450 + $925 = $1375. If you did have the policy, you'd pay two years of premiums (2 × $550 = $1100), all of the $450 claim, and $500 of the $925 claim, for a grand total of $2050.
b. Without the insurance policy, your cost would be $450 + $1200 = $1650. With the policy, you'd pay premiums for two years, and $200 on each claim for a total of 2 × $650 + 2 × $200 = $1700.
c. Without the policy, your cost would be $200 + $1500 = $1700. With the policy, you'd pay premiums for two years, $200 for the first claim, and $1000 for the second claim, for a total of 2 × $300 + $200 + $1000 = $1800.

63. a. You'll pay $25 for each of the office visits, $200 for each trip to the emergency room, and $1000 for the surgery, plus one year of premiums, for a grand total of $25 × 3 + $200 × 2 + $1000 + $350 × 12 = $5675.
b. If you did not have the insurance policy, your cost is the total of the numbers in the second column, which is $8220.

UNIT 4B

QUICK QUIZ

1. **b.** Compound interest always yields a greater balance than simple interest when the APR is the same.

3. **c.** The compound interest formula states $A = P \times (1 + \text{APR})^Y$. With an APR of 6.6%, the balance after five years would be $A = P \times (1 + 0.066)^5$, which means the account increases in value by a factor of 1.066^5.

5. **c.** Compounding interest more often always results in a greater annual percentage yield.

7. **c.** After 20 years, the account earning 10% APR will have grown by a factor of $1.1^{20} = 6.73$, whereas the account earning 5% APR will have grown by a factor of $1.05^{20} = 2.65$. Thus the 10% APR account will have earned almost three times as much interest.

9. **a.** The compound interest formula assumes a constant APR for as many years as you have your money invested in an account.

DOES IT MAKE SENSE?

19. Does not make sense. You earn more interest under compound interest when the APR is the same.

21. Does not make sense. A bank with an APR of 5.9% compounded daily is a better deal than a bank with 6% APR compounded annually (all other things being equal) because the APY in the first case is greater.

23. Makes sense. One's bank account grows based on the APY, and an APR of 5% could certainly result in an APY of 5.1% if the number of compounding periods per year was just right.

BASIC SKILLS & CONCEPTS

25. $x = 12$ (add 3 to both sides).

27. $z = 16$ (add 10 to both sides).

29. $p = 4$ (divide both sides by 3).

31. $z = 4$ (add 1, and divide both sides by 5).

33. $3x - 4 = 2x + 6 \Rightarrow 3x = 2x + 10 \Rightarrow x = 10$.

35. $3a + 4 = 6 + 4a \Rightarrow 3a - 2 = 4a \Rightarrow a = -2$.

37. $6q - 20 = 60 + 4q \Rightarrow 6q = 80 + 4q \Rightarrow 2q = 80 \Rightarrow q = 40$.

39. $t/4 + 5 = 25 \Rightarrow t/4 = 20 \Rightarrow t = 80$.

41. You'll earn 5% of $1000, or $0.05 \times \$1000 = \$50$, each year. After ten years, you will have earned $500 in interest, so that your balance will be $1500.

43. Each year, you'll earn 3% of $3000 = $0.03 \times \$3000 = \$90$. After 20 years, you will have earned $1800 in interest, and your balance will be $4800.

45.

	Yancy		Samantha	
Year	Interest	Balance	Interest	Balance
0	-----	$5000	-----	$5000
1	$250	$5250	$250	$5250
2	$250	$5500	$262.50	$5512.50
3	$250	$5750	$275.63	$5788.13
4	$250	$6000	$289.41	$6077.53
5	$250	$6250	$303.88	$6381.41

Yancy's balance increases by $1250, or 25% of its original value. Samantha's increases by $1,381.41, or 27.6% of its original value. (Note on values shown in the table: the balance in one year added to next year's interest does not necessarily produce next year's balance due to rounding errors. Each value shown is rounded correctly to the nearest cent).

47. $A = \$3000(1 + 0.03)^{10} = \$4031.75$.

49. $A = \$40,000(1 + 0.07)^{25} = \$217,097.31$.

51. $A = \$8000(1 + 0.06)^{25} = \$34,334.97$.

53. $A = \$4000\left(1 + \dfrac{0.035}{12}\right)^{12 \cdot 10} = \$5673.38$.

55. $A = \$15,000\left(1 + \dfrac{0.056}{4}\right)^{4 \cdot 20} = \$45,617.10$.

57. $A = \$2000\left(1 + \dfrac{0.07}{12}\right)^{12 \cdot 15} = \$5697.89$.

59. $A = \$25,000\left(1 + \dfrac{0.062}{4}\right)^{4 \cdot 30} = \$158,318.38$.

61. The APY is the relative increase in the balance of a bank account over the span of one year. The easiest way to compute it when the APR is known is to find the one-year balance, compute the relative increase, and express the answer as a percent. When the principal is not given, any amount can be used – here, we will use $100.

    1-year balance: $A = \$100\left(1 + \dfrac{0.035}{365}\right)^{365 \cdot 1} = \$103.56$.

    The relative increase is $3.56/$100 = 3.56%, so the APY is 3.56%.

63. See #61 for details. The one-year balance is $A = \$100\left(1 + \dfrac{0.0425}{12}\right)^{12 \cdot 1} = \$104.33$. The relative increase (APY) is $4.33/$100 = 4.33%.

65. The one, five, and twenty-year balances are:
    One year: $A = \$3000e^{0.04 \times 1} = \$3122.43$.
    Five years: $A = \$3000e^{0.04 \times 5} = \$3664.21$.
    Twenty years: $A = \$3000e^{0.04 \times 20} = \$6676.62$.
    To compute the APY, find the relative increase in the balance of the account after one year.
    APY = $122.43/$3000 = 4.08%.

67. The one, five, and twenty-year balances are:
    One year: $A = \$10,000e^{0.08 \times 1} = \$10,832.87$.
    Five years: $A = \$10,000e^{0.08 \times 5} = \$14,918.25$.
    Twenty years: $A = \$10,000e^{0.08 \times 20} = \$49,530.32$.
    To compute the APY, find the relative increase in the balance of the account after one year.
    APY = $832.87/$10,000 = 8.33%.

69. The one, five, and twenty-year balances are:
    One year: $A = \$2500e^{0.065 \times 1} = \$2667.90$.
    Five years: $A = \$2500e^{0.065 \times 5} = \$3460.08$.
    Twenty years: $A = \$2500e^{0.065 \times 20} = \$9173.24$.

To compute the APY, find the relative increase in the balance of the account after one year.
APY = $167.90/$2500 = 6.72%.

71. Solve $\$20,000 = P(1 + 0.05)^{10}$ for P to find $P = \dfrac{\$20,000}{(1 + 0.05)^{10}} = \$12,278$.

73. Solve $\$20,000 = P\left(1 + \dfrac{0.06}{12}\right)^{12 \cdot 10}$ for P to find $P = \dfrac{\$20,000}{\left(1 + \dfrac{0.06}{12}\right)^{12 \cdot 10}} = \$10,993$.

75. Solve $\$100,000 = P\left(1 + \dfrac{0.04}{365}\right)^{365 \cdot 18}$ for P to find $P = \dfrac{\$100,000}{\left(1 + \dfrac{0.04}{365}\right)^{365 \cdot 18}} = \$48,677$.

77. Solve $\$100,000 = P\left(1 + \dfrac{0.09}{12}\right)^{12 \cdot 18}$ for P to find $P = \dfrac{\$100,000}{\left(1 + \dfrac{0.09}{12}\right)^{12 \cdot 18}} = \$19,910$.

FURTHER APPLICATIONS

79. After 10 years, Chang has $705.30; after 30 years, he has $1403.40. After 10 years, Kio has $722.52; after 30 years, she has $1508.74. Kio has $17.22, or 2.4% more than Chang after 10 years. She has $105.34, or 7.5% more than Chang after 30 years.

81. To compute the APY, find the one-year balance in the account, using any principal desired ($100 is used here), and then compute the relative increase in the value of the account over one year.
    One-year balance, compounded quarterly:

    $A = \$100\left(1 + \dfrac{0.066}{4}\right)^{4 \cdot 1} = \$106.77$.

    APY = $6.77/$100 = 6.77%.
    One-year balance, compounded monthly:

    $A = \$100\left(1 + \dfrac{0.066}{12}\right)^{12 \cdot 1} = \$106.80$.

    APY = $6.80/$100 = 6.80%.
    One-year balance, compounded daily:

    $A = \$100\left(1 + \dfrac{0.066}{365}\right)^{365 \cdot 1} = \$106.82$.

    APY = $6.82/$100 = 6.82%.
    As the number of compounding periods per year increases, so does the APY, though the rate at which it increases slows down.

83.

Year	Account 1 Interest	Account 1 Balance	Account 2 Interest	Account 2 Balance
0	-----	$1000	-----	$1000
1	$55	$1055	$57	$1057
2	$58	$1113	$60	$1116
3	$61	$1174	$63	$1179
4	$65	$1239	$67	$1246
5	$68	$1307	$70	$1317
6	$72	$1379	$74	$1391
7	$76	$1455	$79	$1470
8	$80	$1535	$83	$1553
9	$84	$1619	$88	$1640
10	$89	$1708	$93	$1733

Account 1 has increased in value by $708, or 70.8%. Account 2 has increased by $733, or 73.3%. (Note on values shown in the table: the balance in one year added to next year's interest does not necessarily produce next year's balance due to rounding errors. Each value shown is rounded correctly to the nearest dollar).

85. After 5 years, Bernard has $A = \$1600(1.04)^5 = \$1946.64$; after 20 years, he has $A = \$1600(1.04)^{20} = \$3505.80$. After 5 years, Carla has $A = \$1400\left(1 + \frac{0.05}{365}\right)^{365 \cdot 5} = \$1797.60$; after 20 years, she has $A = \$1400\left(1 + \frac{0.05}{365}\right)^{365 \cdot 20} = \$3805.33$. Carla has the higher balance after 20 years because her account has a larger APR, and thus the balance catches up to and exceeds Bernard's balance over the long run.

87. For Plan A, solve $\$75,000 = P(1.05)^{35}$ for P to find $P = \frac{\$75,000}{(1.05)^{35}} = \$13,597$. For Plan B, solve $\$75,000 = Pe^{0.045 \times 35}$ for P to find $P = \frac{\$75,000}{e^{0.045 \times 35}} = \$15,526$.

89. To obtain an exact solution for this problem, one must use logarithms (studied in a later chapter). If your initial investment P is to triple, we must solve $3P = P(1.08)^t$ for t in order to find out how long it takes. Begin by dividing both sides by P (it will cancel from both sides), and then take the logarithm of both sides: $\log 3 = \log(1.08)^t$. A property of logarithms says the exponent t can be moved in front of the logarithm, as such: $\log 3 = t \log(1.08)$. Now divide both sides by log 1.08 to arrive at the answer: $t = (\log 3)/(\log 1.08) = 14.3$ years. Thus it will take about 14.3 years for your investment to triple in value. Another way to come to the solution is to try various values of t in

$(1.08)^t$ – you will have found a solution when $(1.08)^t$ is about 3.

91. See exercise 89. Solve $\$100,000 = \$1000(1.07)^t$ for t. Begin by dividing both sides by $1000; this yields $100 = (1.07)^t$. Now take the logarithm of both sides, and apply a property of logarithms to get $\log 100 = t \log 1.07$. Divide both sides by log 1.07 to find $t = (\log 100)/(\log 1.07) = 68.1$ years. Trial and error may also produce the same result if you are patient.

93. The trick to solving this problem is to treat each $500 deposit separately. The first $500 deposit earns 6% interest for 5 full years, so its value is $A = \$500(1.06)^5 = \$669.11$. The $500 deposit made after the first year earns interest for only 4 years, so its value on December 31 of the last year is $A = \$500(1.06)^4 = \$631.24$. In a similar fashion, one can compute the future value of each $500 deposit, realizing that the length of time each is in the bank decreases by one year compared to the previous deposit:

3rd deposit: $A = \$500(1.06)^3 = \$595.51$.

4th deposit: $A = \$500(1.06)^2 = \$561.80$.

5th deposit: $A = \$500(1.06)^1 = \$530$.

The sum of all deposits is the value of the account on December 31 of the last year: $669.11 + $631.24 + $595.51 + $561.80 + $530 = $2987.66.

UNIT 4C

QUICK QUIZ

1. **a.** As the number of compounding periods n increases, so does the APY, and a higher APY means a higher accumulated balance.

3. **c.** The total return on a five-year investment is simply the percent change of the balance over five years.

5. **a.** If you deposit $100 per month for ten years (120 months), you have deposited a total of $12,000 into the savings plan. Since the balance was $22,200, you earned $22,200 – $12,000 = $10,200 in interest.

7. **c.** See table 4.6.

9. **b.** If the bond is selling at 103 points, it is selling at 103% of its face value, which is $1.03 \times \$5000 = \$5150$.

DOES IT MAKE SENSE?

19. Does not make sense. With an APR of 4%, the savings plan formula gives a balance of

$$A = \frac{\$25\left[\left(1+\frac{0.04}{12}\right)^{12\cdot 30}-1\right]}{\left(\frac{0.04}{12}\right)} = \$17,351 \text{ after 30 years,}$$

which probably wouldn't cover expenses for even one year of retirement (and certainly interest earned on that balance would not be enough for retirement). You should plan to save considerably more than \$25 per month for a retirement plan.

21. Does not make sense. Stocks can be a risky investment (you might lose all of your money), and most financial advisors will tell you to diversify your investments.

23. Does not make sense. Stocks are considered a risky investment (especially in the short term). In general, an investment with a greater return is even more risky, and thus the claim that there is no risk at all is dubious, at best.

BASIC SKILLS & CONCEPTS

25. $2^3 = 2 \times 2 \times 2 = 8$.

27. $4^3 = 64$.

29. $16^{1/2} = \sqrt{16} = 4$.

31. $64^{-1/3} = \frac{1}{64^{1/3}} = \frac{1}{\sqrt[3]{64}} = \frac{1}{4}$.

33. $3^4 \div 3^2 = 3^{4-2} = 3^2 = 9$.

35. $25^{1/2} \div 25^{-1/2} = 25^{1/2-(-1/2)} = 25^1 = 25$.

37. $x^2 = 25 \Rightarrow (x^2)^{1/2} = 25^{1/2} \Rightarrow x = \sqrt{25} = 5$. Another solution is $x = -5$.

39. $(x-4)^2 = 36 \Rightarrow ((x-4)^2)^{1/2} = 36^{1/2} \Rightarrow$
    $x - 4 = \sqrt{36} \Rightarrow x = 4 + 6 = 10$. Another solution is $x = -2$.

41. $(t/3)^2 = 16 \Rightarrow ((t/3)^2)^{1/2} = 16^{1/2} \Rightarrow$
    $t/3 = \sqrt{16} \Rightarrow t = 3 \times 4 = 12$. Another solution is $x = -12$.

43. $u^9 = 512 \Rightarrow (u^9)^{1/9} = 512^{1/9} \Rightarrow u = \sqrt[9]{512} = 2$.

45. $A = \frac{\$200\left[\left(1+\frac{0.12}{12}\right)^{12\times 0.75}-1\right]}{\left(\frac{0.12}{12}\right)} = \$1873.71$. Note that

9 months is $3/4 = 0.75$ year, which is why we used $Y = 0.75$ in the formula.

47. $A = \frac{\$600\left[\left(1+\frac{0.06}{12}\right)^{12\times 1.5}-1\right]}{\left(\frac{0.06}{12}\right)} = \$11,271.47$. Note that

18 months is 1.5 years.

49. You save for 40 years, so the value of the IRA is

$$A = \frac{\$75\left[\left(1+\frac{0.05}{12}\right)^{12\times 40}-1\right]}{\left(\frac{0.05}{12}\right)} = \$114,451.51 . \quad \text{Since}$$

you deposit \$75 each month for 40 years, your total deposits are $\frac{\$75}{\text{mo}} \times \frac{12 \text{ mo}}{\text{yr}} \times 40 \text{ yr} = \$36,000$. The value of the account is just over three times the amount you deposited.

51. $A = \frac{\$300\left[\left(1+\frac{0.07}{12}\right)^{12\times 18}-1\right]}{\left(\frac{0.07}{12}\right)} = \$129,216.31 .$ You

have deposited $\frac{\$300}{\text{mo}} \times \frac{12 \text{ mo}}{\text{yr}} \times 18 \text{ yr} = \$64,800$, which is about one-half of the value of the account.

53. Solve $\$75,000 = \frac{PMT\left[\left(1+\frac{0.075}{12}\right)^{12\times 18}-1\right]}{\left(\frac{0.075}{12}\right)}$ for

PMT. $PMT = \frac{\$75,000\left(\frac{0.075}{12}\right)}{\left[\left(1+\frac{0.075}{12}\right)^{12\times 18}-1\right]} = \$164.98$. You

should deposit \$164.98 each month.

55. Solve $\$15,000 = \frac{PMT\left[\left(1+\frac{0.055}{12}\right)^{12\times 3}-1\right]}{\left(\frac{0.055}{12}\right)}$ for PMT.

$PMT = \frac{\$15,000\left(\frac{0.055}{12}\right)}{\left[\left(1+\frac{0.055}{12}\right)^{12\times 3}-1\right]} = \$384.19$. You should

deposit \$384.19 each month.

57. First, find out how large your retirement account needs to be in order to produce \$100,000 in interest (at 6% APR) each year. You want 6% of the total balance to be \$100,000, which means the total balance should be \$100,000/0.06 = \$1,666,667. Now solve

$$\$1,666,667 = \frac{PMT\left[\left(1+\frac{0.06}{12}\right)^{12\times 30}-1\right]}{\left(\frac{0.06}{12}\right)} \text{ for } PMT.$$

$$PMT = \frac{\$1,666,667\left(\dfrac{0.06}{12}\right)}{\left[\left(1+\dfrac{0.06}{12}\right)^{12\times30}-1\right]} = \$1659.18 \text{ . You should}$$

deposit \$1659.18 each month.

59. The total return is the relative change in the investment, and since you invested \$6000 (100 shares at \$60 per share), the total return is $\frac{\$9400-\$6000}{\$6000}=0.567=56.7\%$. The annual return is the APY that would give the same overall growth in five years, and the formula for computing it is $\left(\dfrac{A}{P}\right)^{(1/Y)}-1$. Thus the annual return is $\left(\dfrac{\$9400}{\$6000}\right)^{(1/5)}-1=0.094=9.4\%$.

61. The total return is $\frac{\$11,300-\$6500}{\$6500}=73.8\%$. The annual return is $\left(\dfrac{\$11,300}{\$6500}\right)^{(1/20)}-1=2.8\%$.

63. The total return is $\frac{\$2000-\$3500}{\$3500}=-42.9\%$. The annual return is $\left(\dfrac{\$2000}{\$3500}\right)^{(1/3)}-1=-17.0\%$.

65. The total return is $\frac{\$12,600-\$7500}{\$7500}=68.0\%$. The annual return is $\left(\dfrac{\$12,600}{\$7500}\right)^{(1/10)}-1=5.3\%$.

67. \$500 invested in small-company stocks would be worth $\$500(1+0.126)^{70}=\$2,026,083.50$. \$500 invested in large-company stocks would be worth $\$500(1+0.104)^{70}=\$509,102.40$. For long-term bonds and U.S. Treasury bills, the investment would be worth $\$500(1+0.059)^{70}=\$27,649.50$, and $\$500(1+0.037)^{70}=\$6360.38$, respectively.

69. McDonald's (MCD) had the largest price gain at \$0.31 per share. Since it closed at \$24.42 yesterday, its price two days ago was \$24.42 – \$0.31 = \$24.11.

71. Monsanto is trading at \$21.64 at the close of yesterday, and this is about one dollar less than its 52-week high price of \$22.60. The other stocks are trading at prices significantly lower than their 52-week high.

73. You can expect a dividend payment of \$520, because the stock pays \$0.52 per share in dividends.

75. McDonald's would be the best investment as it has the highest dividend yield (\$0.72 per share).

77. a. Yes, McDonald's earned a profit because its P/E ratio is 12, which means its share price divided by earnings (or profit) per share was 12. This implies share price = 12 × earnings per share.
b. Since share price = 12 × earnings per share, earnings per share = (share price)/12 = \$24.42/12 = \$2.04 per share.
c. The stock prices are about right.

79. a. No, Motorola showed a loss last year (the P/E ratio is listed as "dd").
b. The company didn't earn anything per share as it had a loss last year.
c. Given that the stock is losing money, the stock is expensive.

81. a. Yes, Mueller earned a profit because its P/E ration is 16, which means its share price divided by earnings (or profit) per share was 16. This implies share price = 16 × earnings per share.
b. Since share price = 16 × earnings per share, earnings per share = (share price)/16 = \$27.11/16 = \$1.69 per share.
c. The price is a little expensive (because 16 is larger than the P/E ration of 12 to 14, the historical average).

83. The current yield on a bond is its annual interest payment divided by its current price. A \$1000 bond with a coupon rate of 2% pays \$1000 × 0.02 = \$20 per year in interest, so this bond has a current yield of \$20/\$950 = 2.11%.

85. The annual interest payment on this bond is \$1000 × 0.055 = \$55, so its current yield is \$55/\$1100 = 5%.

87. The price of this bond is 105% × \$1000 = \$1050, and it has a current yield of 3.9%, so the annual interest earned is \$1050 × 0.039 = \$40.95.

89. The price of this bond is 114.3% × \$1000 = \$1143, and its current yield is 6.2%, which means the annual interest will be \$1143 × 0.062 = \$70.87.

91. The 3-year return on SocInvBdA is 10%, which means the fund earned an average of 10% each of the last three years. Thus a \$500 investment is now worth $\$500(1+0.10)^3=\$665.50$.

FURTHER APPLICATIONS

93. After 10 years, the balance in Yolanda's account is

$$A = \frac{\$200\left[\left(1+\dfrac{0.05}{12}\right)^{12\times10}-1\right]}{\left(\dfrac{0.05}{12}\right)} = \$31,056.46 \text{ .} \qquad \text{She}$$

deposits \$200 per month, so her total deposits are $\frac{\$200}{\text{mo}}\times\frac{12\text{ mo}}{\text{yr}}\times10\text{ yr}=\$24,000$. Zach's account is worth $A = \dfrac{\$2400\left[\left(1+\dfrac{0.05}{1}\right)^{1\times10} - 1\right]}{\left(\dfrac{0.05}{1}\right)} = \$30{,}186.94$.

He deposits $2400 per year, so his total deposits are $\dfrac{\$2400}{\text{yr}}\times 10\text{ yr} = \$24{,}000$. Yolanda comes out ahead even though both deposited the same amount of money because the interest in her account is compounded monthly, while Zach's is compounded yearly, which means Yolanda enjoys a higher APY.

95. After 10 years, the balance in Juan's account is

$A = \dfrac{\$400\left[\left(1+\dfrac{0.06}{12}\right)^{12\times10} - 1\right]}{\left(\dfrac{0.06}{12}\right)} = \$65{,}551.74$. His total

deposits are $\dfrac{\$400}{\text{mo}}\times\dfrac{12\text{ mo}}{\text{yr}}\times10\text{ yr} = \$48{,}000$. The

balance in Maria's account is

$A = \dfrac{\$5000\left[(1+0.065)^{10} - 1\right]}{(0.065)} = \$67{,}472.11$. Her total

deposits are $\dfrac{\$5000}{\text{yr}}\times10\text{ yr} = \$50{,}000$. Maria comes

out ahead because she has a higher APR (and APY, it turns out), and she deposits more money over the course of the 10 years.

97. Your balance will be $A = \dfrac{\$50\left[\left(1+\dfrac{0.07}{12}\right)^{12\cdot15} - 1\right]}{\left(\dfrac{0.07}{12}\right)} =$

$15,848.11, so you won't reach your goal.

99. Your balance will be $A = \dfrac{\$100\left[\left(1+\dfrac{0.06}{12}\right)^{12\cdot15} - 1\right]}{\left(\dfrac{0.06}{12}\right)} =$

$29,081.87, so you won't reach your goal.

101. The total return is $\dfrac{\$8.25 - \$6.05}{\$6.05} = 36.4\%$.

Incidentally, this is also the annual return (because you bought the stock a year ago), and it doesn't matter how many shares you bought.

103. The total return was $\dfrac{\$22{,}000{,}000 - \$5000}{\$5000} =$

439,900%. The annual return was

$\left(\dfrac{\$22{,}000{,}000}{\$5000}\right)^{(1/50)} - 1 = 18.3\%$, which is much

higher than the average annual return for large-company stocks.

105. a. At age 35, the balance in Mitch's account is

$A = \dfrac{\$1000\left[(1+0.07)^{10} - 1\right]}{(0.07)} = \$13{,}816.45$. At this

point, it is no longer appropriate to use the savings plan formula as Mitch is not making further deposits into the account. Instead, use this balance as the principal in the compound interest formula, and compute the balance 40 years later.
$A = \$13{,}816.45(1+0.07)^{40} = \$206{,}893.82$. (Note that the rounded balance of $13,816.45 was not used in this computation, but rather the more exact answer stored in a calculator – this is the way it would happen in a banking institution).
b. At age 75, the balance in Bill's account is

$A = \dfrac{\$1000\left[(1+0.07)^{40} - 1\right]}{(0.07)} = \$199{,}635.11$.

c. Mitch deposited $1000 per year for ten years, which is $10,000. Bill deposited $1000 per year for 40 years, which is $40,000, or four times as large as Mitch's total deposits.

UNIT 4D

QUICK QUIZ

1. **a.** Monthly payments go up when the loan principal is larger because you are borrowing more money.
3. **b.** A higher APR means more money goes to paying off interest, and this corresponds with a higher payment.
5. **c.** Every year you'll pay $12,000, and thus after ten years, you'll pay $120,000.
7. **c.** A two-point origination fee just means that you must pay 2% of the loan principal in advance, and 2% of $200,000 is $4000.
9. **c.** Refinancing a loan is not a good idea if you've nearly paid the loan off.

DOES IT MAKE SENSE?

17. Makes sense. For a typical long-term loan, most of the payments go toward interest for much of the loan term. See Figure 4.8 in the text.
19. Does not make sense. Making only the minimum required payment on a credit card is asking for financial trouble as the interest rates on credit cards are typically exorbitant. You'll end up spending a lot of your money on interest.
21. Makes sense. In the worst-case scenario, the ARM loan will be at 6% in the last year you expect to live in the home, so it's certain you'll save money

on the first two years, and you may even save during the last year.

BASIC SKILLS & CONCEPTS

23. a. The starting principal is $80,000 with an annual interest rate of 7%. You'll make 12 payments each year (one per month) for 20 years, and each payment will be $620.

b. Since you make 12 payments per year for 20 years, you'll make $12 \times 20 = 240$ payments in total, which amounts to $240 \times \$620 = \$148,800$ over the term of the loan.

c. Since the loan principal is $80,000, you'll pay $148,800 - \$80,000 = \$68,800$ in interest, and of course $80,000 of your payments will go toward the principal.

25. a. The monthly payment is

$$PMT = \frac{\$50,000\left(\dfrac{0.10}{12}\right)}{1 - \left(1 + \dfrac{0.10}{12}\right)^{-12 \times 20}} = \$482.51 \,.$$

b. The total payment is $\$482.51 \times 12 \times 20 = \$115,802.40$.

c. The money spent on interest is $115,802.40 − $50,000 = \$65,802.40$, while $50,000 goes toward the principal.

27. a. The monthly payment is

$$PMT = \frac{\$200,000\left(\dfrac{0.075}{12}\right)}{1 - \left(1 + \dfrac{0.075}{12}\right)^{-12 \times 30}} = \$1398.43 \,.$$

b. The total payment is $\$1398.43 \times 12 \times 30 = \$503,434.80$.

c. The money spent on interest is $503,434.80 − $200,000 = \$303,434.80$, while $200,000 goes toward the principal.

29. a. The monthly payment is

$$PMT = \frac{\$200,000\left(\dfrac{0.09}{12}\right)}{1 - \left(1 + \dfrac{0.09}{12}\right)^{-12 \times 15}} = \$2028.53 \,.$$

b. The total payment is $\$2028.53 \times 12 \times 15 = \$365,135.40$.

c. The money spent on interest is $365,135.40 − $200,000 = \$165,135.40$, while $200,000 goes toward the principal.

31. a. The monthly payment is

$$PMT = \frac{\$10,000\left(\dfrac{0.12}{12}\right)}{1 - \left(1 + \dfrac{0.12}{12}\right)^{-12 \times 3}} = \$332.14 \,.$$

b. The total payment is $\$332.14 \times 12 \times 3 = \$11,957.04$.

c. The money spent on interest is $11,957.04 − $10,000 = \$1957.04$, while $10,000 goes toward the principal.

33. a. The monthly payment is

$$PMT = \frac{\$150,000\left(\dfrac{0.08}{12}\right)}{1 - \left(1 + \dfrac{0.08}{12}\right)^{-12 \times 15}} = \$1433.48 \,.$$

b. The total payment is $\$1433.48 \times 12 \times 15 = \$258,026.40$.

c. The money spent on interest is $258,026.40 − $150,000 = \$108,026.40$, while $150,000 goes toward the principal.

35. The monthly payment is

$$PMT = \frac{\$150,000\left(\dfrac{0.085}{12}\right)}{1 - \left(1 + \dfrac{0.085}{12}\right)^{-12 \times 30}} = \$1153.37 \,.$$

To compute the interest owed at the end of one month, multiply the beginning principal of $150,000 by the monthly interest rate of 8.5%/12. This gives $150,000(0.085/12) = $1062.50. Because the monthly payment is $1153.37, the amount that goes toward principal is $1153.37 − $1062.50 = $90.87. Subtract this from $150,000 to get a new principal of $150,000 − $90.87 = $149,909.13.

For the second month, repeat this process, but begin with a principal of $149,909.13. The results are shown in the following table through the third month – note that rounding errors at each step propagate through the table, which means that our answers may differ slightly from those that a bank would compute.

End of month	Interest	Payment toward principal	New Principal
1	$1062.50	$90.87	$149,909.13
2	$1061.86	$91.51	$149,817.62
3	$1061.21	$92.16	$149,725.46

37. For the 3-year loan at 7% APR, the monthly payment is $PMT = \dfrac{\$12,000\left(\dfrac{0.07}{12}\right)}{1 - \left(1 + \dfrac{0.07}{12}\right)^{-12 \times 3}} = \$370.53$. The monthly payment for the 4-year loan is $290.15, and for the 5-year loan, it is $243.32 (the computations being similar to those used for the 3-year loan). Since you can afford only the payments on the 5-year loan, that loan best meets your needs.

39. The monthly payment is

$$PMT = \frac{\$5000\left(\frac{0.18}{12}\right)}{1-\left(1+\frac{0.18}{12}\right)^{-12\times1}} = \$458.40.$$

Your total payments are $\$458.40 \times 12 = \$5500.80$.

41. The monthly payment is

$$PMT = \frac{\$5000\left(\frac{0.21}{12}\right)}{1-\left(1+\frac{0.21}{12}\right)^{-12\times3}} = \$188.38.$$

Your total payments are $\$188.38 \times 36 = \$6781.68$.

43. Complete the table as shown in the first month. To find the interest for the second month, compute $\$1093.00 \times 0.015 = \$16.40$. To find the new balance, compute $\$1093.00 + \$75 + \$16.40 - \$200 = \$984.40$. Continue in this fashion to fill out the rest of the table. Note that rounding errors at each step will propagate through the table: a credit card company computing these balances might show slightly different values.

Month	Pmt	Expenses	Interest	New balance
0				$1200.00
1	$200	$75	$18.00	$1093.00
2	$200	$75	$16.40	$984.40
3	$200	$75	$14.77	$874.17
4	$200	$75	$13.11	$762.28
5	$200	$75	$11.43	$648.71
6	$200	$75	$9.73	$533.44
7	$200	$75	$8.00	$416.44
8	$200	$75	$6.25	$297.69
9	$200	$75	$4.47	$177.16
10	$200	$75	$2.66	$54.82

In the 11th month, a partial payment of $\$54.82 + \$54.82 \times 0.015 + \$75 = \$130.64$ will pay off the loan.

45. Complete the table as shown in the first month. To find the interest for the second month, compute $\$179.50 \times 0.015 = \$2.69$. To find next month's balance, compute $\$179.50 + \$150 + \$2.69 - \$150 = \$182.19$. Continue in this fashion to fill out the rest of the table. Note that rounding errors at each step will propagate through the table: a credit card company computing these balances might slow slightly different values.

Month	Pmt	Expenses	Interest	Balance
0				$300.00
1	$300	$175	$4.50	$179.50
2	$150	$150	$2.69	$182.19
3	$400	$350	$2.73	$134.92
4	$500	$450	$2.02	$86.94
5	$0	$100	$1.30	$188.24
6	$100	$100	$2.82	$191.06
7	$200	$150	$2.87	$143.93
8	$100	$80	$2.16	$126.09

47. Under Option 1, the monthly payment is

$$PMT = \frac{\$200,000\left(\frac{0.08}{12}\right)}{1-\left(1+\frac{0.08}{12}\right)^{-12\times30}} = \$1467.53,$$ and the total

cost of the loan is $\$1467.53 \times 12 \times 30 = \$528,310.80$. Under Option 2, the monthly

payment is $PMT = \dfrac{\$200,000\left(\frac{0.075}{12}\right)}{1-\left(1+\frac{0.075}{12}\right)^{-12\times15}} = \$1854.02,$

and the total cost of the loan is $\$1854.02 \times 12 \times 15 = \$333,723.60$. You'll pay considerably more interest over the term of the loan in Option 1, but you may be able to afford the payments under that option more easily.

49. Under Option 1, the monthly payment is

$$PMT = \frac{\$60,000\left(\frac{0.0715}{12}\right)}{1-\left(1+\frac{0.0715}{12}\right)^{-12\times30}} = \$405.24,$$ and the total

cost of the loan is $\$405.24 \times 12 \times 30 = \$145,886.40$. Under Option 2, the monthly

payment is $PMT = \dfrac{\$60,000\left(\frac{0.0675}{12}\right)}{1-\left(1+\frac{0.0675}{12}\right)^{-12\times15}} = \$530.95,$

and the total cost of the loan is $\$530.95 \times 12 \times 15 = \$95,571.00$. See #47 for pros and cons.

51. Under Choice 1, your monthly payment is

$$PMT = \frac{\$120,000\left(\frac{0.08}{12}\right)}{1-\left(1+\frac{0.08}{12}\right)^{-12\times30}} = \$880.52,$$ and the

closing costs are $1200. For Choice 2, the

payment is $PMT = \dfrac{\$120{,}000\left(\dfrac{0.075}{12}\right)}{1-\left(1+\dfrac{0.075}{12}\right)^{-12\times30}} = \$839.06$,

and the closing costs are $\$1200 + \$120{,}000(0.02) = \$3600$. You'll save $\$880.52 - \$839.06 = \$41.46$ each month with Choice 2, though it will take $\$2400/(\$41.46$ per month$) = 58$ months (about 5 years) to recoup the higher closing costs, and thus Choice 2 is the better option only if you intend to keep the house for at least 5 years.

53. Under Choice 1, your monthly payment is

$PMT = \dfrac{\$120{,}000\left(\dfrac{0.0725}{12}\right)}{1-\left(1+\dfrac{0.0725}{12}\right)^{-12\times30}} = \$818.61$, and the

closing costs are $\$1200 + \$120{,}000(0.01) = \$2400$. For Choice 2, the payment is

$PMT = \dfrac{\$120{,}000\left(\dfrac{0.0675}{12}\right)}{1-\left(1+\dfrac{0.0675}{12}\right)^{-12\times30}} = \$778.32$, and the

closing costs are $\$1200 + \$120{,}000(0.03) = \$4800$. You'll save $\$818.61 - \$778.32 = \$40.29$ each month with Choice 2, though it will take $\$2400/(\$40.29$ per month$) = 60$ months (5 years) to recoup the higher closing costs, and thus Choice 2 is the better option only if you intend to keep the house for at least 5 years.

55. a. The monthly payment is

$PMT = \dfrac{\$30{,}000\left(\dfrac{0.09}{12}\right)}{1-\left(1+\dfrac{0.09}{12}\right)^{-12\times20}} = \$269.92$.

b. The monthly payment is

$PMT = \dfrac{\$30{,}000\left(\dfrac{0.09}{12}\right)}{1-\left(1+\dfrac{0.09}{12}\right)^{-12\times10}} = \$380.03$.

c. If you pay the loan off in 20 years, the total payments will be $\$269.92 \times 12 \times 20 = \$64{,}780.80$. If you pay it off in 10 years, the total payments will be $\$380.03 \times 12 \times 10 = \$45{,}603.60$.

57. Following example 10 in the text, the interest on the $\$150{,}000$ loan in the first year will be approximately $7\% \times \$150{,}000 = \$10{,}500$, which means the monthly payment will be about $\$10{,}500/12 = \$875$. With the 5% ARM, the interest will be approximately $5\% \times \$150{,}000 = \$7500$, and thus the monthly payment will be about $\$7500/12 = \$625$. You'll save $\$875 - \$625 = \$250$ each month with the ARM. In the third

year, the rate on the ARM will be 1.5% higher than the fixed rate loan, and thus the yearly interest payments will differ by $\$150{,}000 \times 0.015 = \$2250$, which amounts to $\$187.50$ per month.

FURTHER APPLICATIONS

59. Solve

$\$500 = \dfrac{P\left(\dfrac{0.09}{12}\right)}{1-\left(1+\dfrac{0.09}{12}\right)^{-12\times30}} = P\dfrac{\left(\dfrac{0.09}{12}\right)}{1-\left(1+\dfrac{0.09}{12}\right)^{-12\times30}}$ for

P by computing the fraction on the right, and dividing it into \$500. This results in P = \$62,141 (rounded to the nearest dollar), and it represents the largest loan you can afford with monthly payments of \$500. The loan will pay for 80% of the house you wish to buy (you must come up with a 20% down payment). Thus 80% × (*house price*) = \$62,141, which means *house price* = \$62,141/0.80 = \$77,676. You can afford a house that will cost about \$78,000.

61. a. The monthly payment for the first loan is

$PMT = \dfrac{\$10{,}000\left(\dfrac{0.08}{12}\right)}{1-\left(1+\dfrac{0.08}{12}\right)^{-12\times15}} = \$95.57$. The payments

for the other two loans are calculated similarly; you should get \$130.17 and \$158.34, respectively.
b. For the first loan, you'll pay $\$95.57 \times 12 \times 15 = \$17{,}202.60$ over its term. The total payments for the other loans are computed in a similar fashion – you should get \$31,240.80 and \$19,000.80, respectively. Add them together to get \$67,444.20.
c. The monthly payment for the consolidated loan

will be $PMT = \dfrac{\$37{,}500\left(\dfrac{0.085}{12}\right)}{1-\left(1+\dfrac{0.085}{12}\right)^{-12\times20}} = \$325.43$, and

the total payment over 20 years will be \$78,103.20. You'll pay about \$11,000 more for the consolidated loan ($\$78{,}103.20 - \$67{,}444.20 = \$10{,}659$), but your monthly payments will be lower for the first ten years (\$325.43 versus $\$95.57 + \$130.17 + \$158.34 = \$384.08$). If, as a recent graduate, your budget is tight, it may be worth it to consolidate the loans. It also means you'll have to keep track of only one loan instead of three, though with automatic withdrawals offered by most banks, this isn't as much of an issue today as it might have been 15 years ago.

63. a. The annual ($n = 1$) payment is \$8718.46; the monthly ($n = 12$) payment is \$716.43, the bi-monthly ($n = 26$) payment is \$330.43, and the weekly ($n = 52$) payment is \$165.17. The weekly

payment is computed below; the others are done in a similar fashion.

$$PMT = \frac{\$100,000\left(\dfrac{0.06}{52}\right)}{1-\left(1+\dfrac{0.06}{52}\right)^{-52\times20}} = \$165.17 \ .$$

b. The total payout for each scenario is as follows: $174,369.20 ($n = 1$); $171,943.20 ($n = 12$); $171,823.60 ($n = 26$); and $171,776.80. The total payout when $n = 52$ is shown below; the others are done in a similar fashion.

$$\frac{\$165.17}{\text{week}} \times \frac{52 \text{ weeks}}{\text{year}} \times 20 \text{ years} = \$171,776.80 \ .$$

c. The total payout decreases as n increases.

UNIT 4E

QUICK QUIZ

1. **a**. Gross income is defined as the total of all income you receive.
3. **a**. A tax credit reduces your tax bill by the dollar amount of the credit, so your tax bill will be reduced by $1000.
5. **b**. You can claim deductions of $4000 + $2000 = $6000 as long as you itemize deductions.
7. **c**. FICA taxes are taxes levied on income from wages (and tips) to fund the Social Security and Medicare programs.
9. **b**. Assuming all are of the same filing status, Jerome pays the most because his FICA taxes will be highest, followed by Jenny, and then Jacqueline.

DOES IT MAKE SENSE?

21. Does not make sense. We know nothing of the rest of the picture for these two individuals – they may have very different deductions, and thus different tax bills.
23. Makes sense. You can deduct interest paid on the mortgage of a home if you itemize deductions, and for many people, the mortgage interest deduction is the largest of all deductions.
25. Makes sense (in some scenarios). Bob and Sue might be hit with the "marriage penalty" when they file taxes, and if they postpone their wedding until the new year, they could avoid the penalty for at least the previous tax year.
27. Makes sense. Those who are self-employed must pay both the employee and employer ends of the FICA tax, and for any profit up to $94,200, the rate is 7.65%. Self-employed people must pay twice that rate, or 15.3%.

BASIC SKILLS & CONCEPTS

29. Antonio's gross income was $47,200 + $2400 = $49,600. His AGI was $49,600 – $3500 = $46,100. His taxable income was $46,100 – $3300 – $5150 = $37,650.
31. Isabella's gross income was $88,750 + $4900 = $93,650. Her AGI was $93,650 – $6200 = $87,450. Her taxable income was $87,450 – $3300 – $9050 = $75,100.
33. Your itemized deductions total to $8600 + $2700 + $645 = $11,945, and since this is larger than the standard deduction of $10,300, you should itemize your deductions.
35. Suzanne's gross income is $33,200 + $350 = $33,550, her AGI is $33,550 – $500 = $33,050, and her taxable income is $33,050 – $3300 – $5150 = $24,600. She should claim the standard deduction because it is much higher than her itemized deduction of $450.
37. Wanda's gross income was $35,400 + $500 = $35,900, her AGI was the same, and her taxable income was $35,900 – $3300 × 3 – $5150 = $20,850. She should claim the standard deduction because it is higher than her itemized deduction of $1500.
39. Gene owes 10% × ($7550) + 15% × ($30,650 – $7550) + 25% × ($35,400 – $30,650) = $5408 (taxes are rounded to the nearest dollar).
41. Bobbi owes 10% × ($7550) + 15% × ($30,650 – $7550) + 25% × ($61,850 – $30,650) + 28% × ($77,300 – $61,850) = $16,346.
43. Paul owes 10% × ($10,750) + 15% × ($41,050 – $10,750) + 25% × ($89,300 – $41,050) – $1000 = $16,683.
45. Winona and Jim owe 10% × ($15,100) + 15% × ($61,300 – $15,100) + 25% × ($105,500 – $61,300) – $2000 = $17,490.
47. Their bill will be reduced by $500.
49. Her taxes will not be affected because she is claiming the standard deduction, and a $1000 charitable contribution must be itemized.
51. His tax bill will be reduced by 28% of $1000, or $280, provided he is far enough into the 28% bracket that his $1000 contribution does not drop him into the lower bracket.
53. The $1800 per month of your payment that goes toward interest would be deductible, which means you'd save 33% × $1800 = $594 each month in taxes. Thus your $2000 mortgage payment would effectively be only $2000 – $594 = $1406, which is less than your rent payment of $1600. Buying the house is cheaper. This solution assumes that the $1800 × 12 = $21,600 interest deduction would not drop you into the 28% bracket (if it did, the solution would be more complicated).

55. Maria saves 33% of $10,000, or $3300 in taxes, which means the true cost of her mortgage interest is $6700. Steve saves 15% of $10,000, or $1500 in taxes, which means the true cost of his mortgage interest is $8500. This solution assumes their deductions do not drop them into the lower tax bracket.

57. Luis will pay 7.65% × $28,000 = $2142 in FICA taxes. His taxable income is $28,000 − $3300 − $5150 − $2500 = $17,050. He will pay 10% × ($7550) + 15% × ($17,050 − $7550) = $2180 in income taxes, and thus his total tax bill will be $2142 + $2180 = $4322. This is $4322/$28,000 = 15.4% of his gross income, so his effective tax rate is 15.4%.

59. Jack will pay 7.65% × $44,800 = $3427 in FICA taxes. His gross income is $44,800 + $1250 = $46,050, and his taxable income is $46,050 − $3300 − $5150 − $2000 = $35,600. He will pay 10% × ($7550) + 15% × ($30,650 − $7550) + 25% × ($35,600 − $30,650) = $5458 in income taxes, and thus his total tax bill will be $3427 + $5458 = $8885. This is $8885/$46,050 = 19.3% of his gross income, so his effective tax rate is 19.3%.

61. Brittany will pay 7.65% × $48,200 = $3687 in FICA taxes. Her taxable income is $48,200 − $3300 − $5150 = $39,750. She will pay 10% × ($7550) + 15% × ($30,650 − $7550) + 25% × ($39,750 − $30,650) = $6495 in income taxes, and thus her total tax bill will be $3687 + $6495 = $10,182. This is $10,182/$48,200 = 21.1% of her gross income, so her effective tax rate is 21.1%.

63. Pierre will pay 7.65% × ($94,200) + 1.45% × ($120,000 − $94,200) = $7580 in FICA taxes, and because his taxable income is $120,000 − $3300 − $5150 = $111,550, he will pay 10% × ($7550) + 15% × ($30,650 − $7550) + 25% × ($74,200 − $30,650) + 28% × ($111,550 − $74,200) = $25,566 in income taxes. His total tax bill is $7580 + $25,566 = $33,146, so his overall tax rate is $33,146/$120,000 = 27.6%. Katarina will pay nothing in FICA taxes, and her income taxes will be 5% × ($30,650) + 15% × ($111,550 − $30,650) = $13,668 – like Pierre, her taxable income is $111,550, but she is taxed at the special rates for dividends and long-term capital gains. Her overall tax rate is $13,668/$120,000 = 11.4%.

65. Because you are in the 15% tax bracket, every time you make a $400 contribution to your tax-deferred savings plan, you save 15% × $400 = $60 in taxes. Thus your take-home pay is reduced not by $400, but by $340.

67. Because you are in the 25% tax bracket, every time you make an $800 contribution to your tax-deferred savings plan, you save 25% × $800 = $200 in taxes. Thus your take-home pay is reduced not by $800, but by $600.

FURTHER APPLICATIONS

69. Gabriella's taxable income is $44,500 − $3300 − $5150 = $36,050, so her income tax is 10% × ($7550) + 15% × ($30,650 − $7550) + 25% × ($36,050 − $30,650) = $5570. Roberto's taxable income is $33,400 − $3300 − $5150 = $24,950, so his income tax is 10% × ($7550) + 15% × ($24,950 − $7550) = $3365. If they delay their marriage, they will pay $5570 + $3365 = $8935 in taxes. If they marry, their combined AGI will be $44,500 + $33,400 = $77,900, and their taxable income will be $77,900 − $3300 × 2 − $10,300 = $61,000. Their income taxes will be 10% × ($15,100) + 15% × ($61,000 − $15,100) = $8395, which is lower than the taxes they would pay as individuals, and thus they will not face a marriage penalty (in fact, they'll benefit from becoming married).

71. Steve's taxable income is $185,000 − $3300 − $5150 = $176,550, so his income tax is 10% × ($7550) + 15% × ($30,650 − $7550) + 25% × ($74,200 − $30,650) + 28% × ($154,800 − $74,200) + 33% × ($176,550 − $154,800) = $44,853. Mia's tax is, of course, the same, so if they delay their marriage, they will pay $89,706. If they marry, their combined AGI will be $185,000 × 2 = $370,000, and their taxable income will be $370,000 − $3300 × 2 − $10,300 = $353,100. Their income taxes will be 10% × ($15,100) + 15% × (61,300 − 15,100) + 25% × ($123,700 − $61,300) + 28% × ($188,450 − $123,700) + 33% × ($336,550 − $188,450) + 35% × ($353,100 − $336,550) = $96,836, which is more than the taxes they would pay as individuals, and thus they will face a marriage penalty.

UNIT 4F

QUICK QUIZ

1. **b.** Bigprofit.com had more outlays (expenses) than receipts (income) by $1 million, so it ran a deficit for 2006 of $1 million. Its debt is the total amount owed to lenders over the years; this is $7 million.

3. **c.** The error in predicting the budget is around 10%, and 10% of $2.5 trillion is $250 billion. Thus if receipts were 10% below what was predicted, the deficit would be $350 billion, and if receipts were 10% above what was predicted, there would be a surplus of $150 billion.

5. **a.** Interest on the debt must be paid so that the government does not default on its loans, and Medicare is an entitlement program that is part of mandatory spending. On the other hand, all money spent on defense is considered discretionary spending.

7. **b.** The $100 billion is supposed to be deposited into an account used to pay off future Social Security recipients, but instead, the government spends the money at hand, and essentially writes IOUs to itself, and places these in the account.

9. **b.** The gross debt is the sum of all of the money that the government owes to individuals who have purchased bonds, and the money it owes to itself.

DOES IT MAKE SENSE?

19. Makes sense. Each U.S. citizen would have to spend about $30,000 to retire the federal debt, and it's certainly true that this exceeds the personal debt of many people.

21. Does not make sense. The financial health of the government is very much dependent upon what happens with Social Security and the FICA taxes collected to fund it. The term "off-budget" is nothing more than a label that separates the Social Security program from the rest of the budget.

23. Does not make sense. The government has a difficult time predicting deficits or surpluses even one year out – it cannot predict what will happen with any level of accuracy in ten years.

BASIC SKILLS & CONCEPTS

25. **a.** Your receipts are $38,000, and your outlays are $12,000 + $6000 + $1200 + $8500 = $27,700, so you have a surplus.
    b. A 3% raise means your receipts will be $38,000 + 3% × $38,000 = $39,140, and your outlays will be $27,700 + $8500 = $36,200, so you'll still have a surplus, though it will be smaller.
    c. Your receipts will be $39,140 (see part **b**), and your outlays will be $27,700 × 1.01 + $7500 = $35,477, so you will be able to afford it.

27. $\dfrac{\$9 \times 10^{12}}{170 \times 10^6 \text{ workers}} = \text{about } \dfrac{\$53,000}{\text{worker}}$.

29. **a.** The interest on the debt will be 8.2% × $773,000 = $63,000 (rounded to the nearest thousand dollars).
    b. The total outlays are $600,000 + $200,000 + $250,000 + $63,000 = $1,113,000, so the year-end

deficit is $1,050,000 − $1,113,000 = −$63,000, and the year-end accumulated debt is −$773,000 − $63,000 = −$836,000.
    c. The interest on the debt in 2009 will be 8.2% × $836,000 = $69,000 (rounded to the nearest thousand dollars).
    d. The total outlays are $600,000 + $200,000 + $69,000 = $869,000, so the year-end surplus is $1,100,000 − $869,000 = $231,000, and the year-end accumulated debt is −$836,000 + $231,000 = −$605,000, assuming all surplus is devoted to paying down the debt.

31. With a 1% decrease in total receipts in 2006, total receipts would be 99% × $2407 = $2383, and thus the deficit would be $2383 − $2654 = −$271 (billion). A 0.5% increase in total outlays would result in total outlays of 100.5% × $2654 = $2667, and thus the deficit would be $2407 − $2667 = −$260 (billion).

33. Individual income taxes were 44% of $2407 billion = $1059 billion, or just over $1 trillion.

35. Excise taxes were 3% of $2407 billion = $72 billion.

37. Medicare expenses were 13% of $2654 = $345 billion.

39. Because *unified net income – off-budget net income = on-budget net income*, we have $40 billion – $180 billion = –$140 billion. In other words, despite the fact that the government proclaimed a surplus of $40 billion for this year, it actually ran a deficit of $140 billion due to the fact that the money earmarked for the Social Security trust fund was never deposited there (it was spent on other programs).

41. The government could cut discretionary spending, it could borrow money by issuing Treasury notes, or it could raise taxes (of course, a combination of these three would also be an option).

FURTHER APPLICATIONS

43. $\$9 \times 10^{12} \times \dfrac{1\text{ s}}{\$1} \times \dfrac{1\text{ hr}}{3600\text{ s}} \times \dfrac{1\text{ d}}{24\text{ hr}} \times \dfrac{1\text{ yr}}{365\text{ d}} = 285{,}388 \text{ yr}$.
    It would take about 285,000 years.

45. In ten years, the debt will be
    $$\$9 \times 10^{12}(1 + 0.01)^{10} = \$9.9 \times 10^{12},$$
    or about $9.9 trillion. In 50 years, it will be
    $$\$9 \times 10^{12}(1 + 0.01)^{50} = \$14.8 \times 10^{12},$$
    or about $14.8 trillion.

47. The projected outlays are $2590 billion + $223 billion = $2813 billion. If the outlays were higher by 5%, they would be 105% × $2813 billion = $2954 billion, and if the receipts were lower by 5%, they would be 95% × $2590 billion = $2461 billion. This would result in a deficit of $2461 billion − $2954 billion = −$493 billion.

49. $PMT = \dfrac{\$5 \times 10^{12} \left(\dfrac{0.04}{1} \right)}{1 - \left(1 + \dfrac{0.04}{1} \right)^{-1 \times 10}} = \$6.16 \times 10^{11}$, which is

$616 billion per year.

51. $\$9 \times 10^{12} \times \dfrac{1 \text{ wk}}{\$150 \times 10^6} \times \dfrac{1 \text{ yr}}{52 \text{ wk}} = 1154 \text{ years}$.

UNIT 5A

QUICK QUIZ

1. **a.** The population is the complete set of people or things that are being studied.

3. **a.** A representative sample is one where the characteristics of the sample match those of the population.

5. **c.** The experiment is not blind, because the participants know whether they are receiving money or not, and the teacher knows to which group each student belongs.

7. **c.** A placebo effect results when participants of the study improve because they think they are receiving the treatment.

9. **b.** With a margin of error of 3%, both Poll X and Poll Y have predictions that overlap (in that Poll X predicts Powell will receive 46% to 52% of the vote, and Poll Y predicts he will receive 50% to 56% of the vote), and so they are consistent in their predictions.

DOES IT MAKE SENSE?

19. Does not make sense. A sample is a subset of the population, and thus it cannot exceed the size of the population.

21. Does not make sense. The control group should receive the same treatment given to the treatment group.

23. Makes sense. The poll makes a prediction about the outcome of the election, and though we can place a reasonable level of confidence in the results of the poll, it cannot tell the future for us.

BASIC SKILLS & CONCEPTS

25. The population is all Americans, and the sample was the set of those surveyed. The population parameters are the opinions of Americans on Iran, and the sample statistics consist of the opinions of those who were surveyed.

27. The population is the set of *USA Today* readers (or perhaps just those who read *USA Today* on the Internet), and the sample is the set of people who responded to the survey. The population parameter is the percent of *USA Today* readers who drink at least one caffeinated beverage per day, and the sample statistic is the percent of those who responded to the survey that drink at least one such beverage per day.

29. The population is the set of adult Americans, and the sample is the 2435 adults who were sampled. The population parameters are the opinions of American adults on public schools, whereas the sample statistics are the opinions of those surveyed.

31. *Step 1*: The goal is to determine the average time spent per day listening to iPods among a population of students at a middle school. *Step 2*: Randomly select (from a list of all students at the school) a sample from the population. *Step 3*: Determine iPod use from those in the sample, either with an interview, a survey, or a device that measures more precisely when the iPod is in use. *Step 4*: Conclude that the iPod use among all students is similar to iPod use among the sample. *Step 5*: Assess the results and formulate a conclusion.

33. *Step 1*: The population of this study is American male college students, and the goal is to determine the percentage of the population that works at Sudoku puzzles at least once per week. *Step 2*: Choose a representative sample of male college students. *Step 3*: Survey students and determine the percentage of those in the sample who do Sudoku puzzles. *Step 4*: Infer the percentage of the population who do Sudoku puzzles from the results of the study. *Step 5*: Assess the results and draw conclusions.

35. *Step 1*: The population in the study is all cars made in Japan, and the goal is to find the average lifetime of windshield wipers on those cars. *Step 2*: Select a representative sample of cars made in Japan. *Step 3*: Formulate a method to determine lifetimes of windshield wipers, and compute the average lifetime. *Step 4*: Infer the average lifetime found in *Step 3* is the same for all cars made in Japan. *Step 5*: Assess the results and formulate a conclusion.

37. The first 50 girls met in the cafeteria is the most representative sample for girls at a particular school because they are a random sample of students at that school. The other groups are not likely to be representative for the following reasons: athletes are generally required to carry a minimum GPA in order to participate in a sports program; those in advanced placement courses are typically good students with above average study habits; and students who participate in theater are more likely to have better time management skills than a random student. All of these would contribute to a bias in the samples regarding study habits.

39. This is an example of stratified sampling as the sample of taxpayers is broken in several categories (or strata).

40. This is an example of convenience sampling, because the sample is the set of people who voluntarily respond to a survey.

41. This is an example of stratified sampling as the participants are divided into groups based on age.

43. This is a simple random sampling because the people are selected at random.

45. This is an observational study, with a case-control element, where the cases are those who have a tendency to lie, and the controls are those who do not lie.

47. This is an observational study, with no case-control component.

49. This is an experiment where the treated group consists of swimmers who included weight training in their workouts, and the control group consists of those who did no weight training. It would be difficult to incorporate the ideas of placebos and blinding in this experiment as there is no way to mask the fact that the subject is weight training.

51. An observational study would be sufficient to determine the average number of hours worked.

53. Use an observational study with a case-control to find the answer. The cases would be teenagers with a diet high in dairy products, while the control would be those who have a diet not so dependent upon dairy products.

55. An experiment would answer this question best. Recruit people to participate in the study, and ask one group to take a multi-vitamin every day, while the other group receives a placebo. Observe the rate of strokes among each group to determine the effect of multi-vitamins.

57. The confidence interval is 50.5% to 55.5%, and because there are only two candidates, it is likely (though not guaranteed) the Republican will win, so a victory party is called for.

59. The confidence interval is 49% to 67%. You cannot be sure a majority of Americans changed vacation plans because it's possible (49%) that a minority changed plans.

FURTHER APPLICATIONS

61. Because a significantly larger percentage of those who received the drug showed improvement compared to those who did not receive the drug, there is good evidence that the treatment is effective (or at least more effective than no treatment at all).

63. Because a significantly larger percentage of those who received the drug showed improvement compared to those who did not receive the drug, there is good evidence that the treatment is effective.

65. a. The population is adults with children (or perhaps people who think they may have children in the future), and the population parameter is the percentage of adults who responded yes.

b. The sample is the 748 adults, and the sample statistic is the percentage of those adults who responded yes.

c. The confidence interval is 59.3% to 66.7%.

67. a. The population is all people of working age, and the population parameter is the percentage of the population that is unemployed.

b. The sample is people of working age in 60,000 households, and the sample statistic is the percentage of those in the sample that are unemployed.

c. The confidence interval is 6.2% to 6.6%.

69. a. The population is registered voters, and the population parameters are the percentage of voters who think the government is listening to your phone conversations (yes), and the percentage who think the government is not listening (no).

b. The sample is the 900 voters surveyed, and the sample statistics are the percentage of those surveyed who think the government is listening to their phone conversations, and the percentage who think the government is not listening.

c. The confidence interval for those responding "no" is 27% to 33%, and for those responding "yes," it is 55% to 61%.

UNIT 5B

QUICK QUIZ

1. **c.** When the wrong technique is used in a statistical study, one should put little faith in the results.

3. **a.** The fact that the researchers interviewed only those living in dormitories means the sample chosen may not be representative of the population of all freshmen, some of whom don't live on campus.

5. **b.** The quantities that a statistical study attempts to measure are called variables of interest, and this study is measuring the weights of 6-year-olds.

7. **c.** It's reasonable to assume that people who use sunscreen do so because they spend time in the sun, and this can lead to sunburns.

9. **b.** A self-selected survey suffers from participation bias, where people make the choice to participate (that is, they select themselves, rather than being randomly selected by a polling company).

DOES IT MAKE SENSE?

15. Does not make sense. A TV survey with phone-in responses suffers from participation bias, whereas a survey carried out by professional pollsters more likely than not will give better results.

17. Does not make sense. Statistical studies can rarely (if ever) make the claim that the results are proven beyond all doubt.

BASIC SKILLS & CONCEPTS

19. You should not put much faith in the study because the sample is not likely to be representative of the population.

21. A study done by a conservative group to determine opinions about a Democratic tax plan could easily include biases, so you should doubt the results.

23. You should doubt a poll that suffers from participation bias, as this one does.

25. It is very difficult to quantify or even define a fulfilling life, and thus this study is essentially meaningless (and you should not trust it).

27. You should doubt the results of this study because there is no control group to which results might be compared.

29. There are at least two reasons to doubt the study: it suffers from the availability error, and the sample may not be representative of the population (unless the population is defined to be those who shop at that store).

31. There are plenty of reasons to doubt this claim, among them is that the sample (long-distance runners) is not representative of the population (all athletes).

33. The Census Bureau is a reputable organization, and there's no reason to doubt the claim (one might argue that a Census statistic which has far reaching political implications could be biased, but this claim is pretty neutral).

35. There's no reason to doubt this claim.

FURTHER APPLICATIONS

37. People who visit the Web site are more inclined toward saving animals rather then getting rid of them, and thus the vote on the issue of euthanasia is heavily biased (this is an example of participation bias).

39. A potential selection bias exists in this survey because the people who vote from 7:00 to 7:30 a.m. may not be a representative sample of the voting population (those in the sample are more likely to be working people who vote on the way to their jobs).

41. This survey will suffer from a selection bias due to the fact that the National Guard is charged with protecting the borders, and thus their views on immigration won't be representative of the entire country.

43. Planned Parenthood members have different views on the issue of contraceptives for high school

students than the population of American adults (that is, they are not a representative sample), and thus this survey is influenced by the selection bias.

45. Answers will vary.

47. The stat-byte makes no mention of how "hours of delay are measured," nor are there any comparative values offered that would put the "27,144 hours" into context.

49. The way in which the question was asked may have influenced responses, and we know nothing of the sample or its size.

51. We don't know what question was asked to obtain the responses (was it the same question?), and we don't know the sample or its size.

53. A headline like this implies "illegal drugs" to most readers, and the government study includes smoking and drinking (both of which are legal and largely acceptable practices in many segments of society). The study is based upon top movie rentals, and yet the headline implies all movies. It's a misleading headline.

55. The headline might have some truth to it (and to the credit of the headline writer, the word "may" is included in the claim), but the study is based upon a small sample, and there's a potential conflict of interest as Welch Foods (think Welch Grape Juice) has an interest in a favorable outcome for the study.

UNIT 5C

QUICK QUIZ

1. **b.** The relative frequency of a category expresses its frequency as a fraction or percentage of the total, and thus is $25/100 = 0.25$.

3. **b.** Qualitative data describe non-numerical categories.

5. **a.** Depending on how you want to display the data, you could actually use any of the three options listed. However, a line chart is used for numerical data, which means you would have to suppress the names of the tourist attractions (your categories would becomes bins, such as "0-2 million"). For a pie chart, you would need to compute the relative frequencies of the data, and while that's not terribly difficult, it requires an extra step. It would be most appropriate to use a bar graph, with the various tourist attractions as categories (e.g. Disney World, Yosemite National Park, etc.), and the annual number of visitors plotted on the y-axis.

7. **a.** With 100 data points, each precise to the nearest 0.001, it would be best to bin the data into several categories.

9. **c.** A histogram is a bar graph for quantitative data categories.

DOES IT MAKE SENSE?

17. Does not make sense. A frequency table has two columns, one of which lists the frequency of the various categories.

19. Does not make sense. The width of the bars in a bar chart is not important (though they should all be of the *same* width in a histogram, and even for those bar charts that are not histograms, even-width bars look better).

21. Does not make sense. The position of the wedges in a pie chart is immaterial.

23. Does not make sense. Bar charts can be used for qualitative data: the categories are the qualities of concern, and the vertical axis shows the frequency of each category (or the relative frequencies).

BASIC SKILLS & CONCEPTS

25.

Grade	Freq.	Rel. freq.	Cum. freq.
A	2	0.10	2
B	5	0.25	7
C	8	0.40	15
D	3	0.15	18
F	2	0.10	20
Total	20	1.00	20

27. Hair color is qualitative data because it describes a non-numerical quality of hair.

29. The taste of sausage is qualitative data because it describes a non-numerical quality of sausage (the numbers used are really just labels for the taste).

31. The responses to the question are qualitative data because they are non-numerical responses.

33. The dessert selections are qualitative data because they are non-numerical choices in the poll.

35.

Bin	Freq.	Rel. freq.	Cum. freq.
95–99	3	0.15	3
90–94	2	0.10	5
85–89	3	0.15	8
80–84	2	0.10	10
75–79	4	0.20	14
70–74	1	0.05	15
65–69	4	0.20	19
60–64	1	0.05	20
Total	20	1.00	20

37.

39.

41.

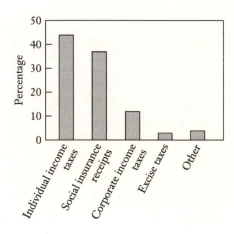

43.

Bin	Frequency
20–29	0
30–39	12
40–49	13
50–59	5
60–69	3
70–79	1
Total	34

45. Answers will vary.

FURTHER APPLICATIONS

47. a. The ages are quantitative categories.
b.

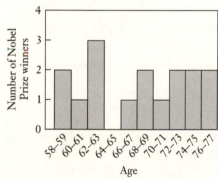

c. People who win the Nobel Prize (in literature) do so rather late in their career, and the data are evenly distributed (this would be more apparent if half as many bins were used).

49. a. The ethnic groups are qualitative categories.
b.

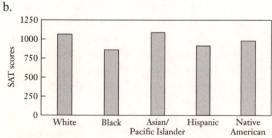

c. With the exception of Asian/Pacific Islanders, minority students don't do as well on the SAT as White students.

51. a. The land masses are qualitative categories.
b.

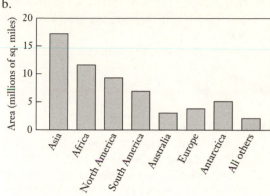

c. Together, Asia and Africa make up almost one-half of the worlds land masses.

53. a. The religions are qualitative categories.
b.

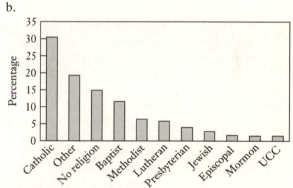

c. Catholics outnumber every category by a significant percentage.

55. a. The years are quantitative categories.
b.

c. Families are getting smaller in the U.S.

57. a.

b. President 35 (Kennedy) could claim to be the youngest.

c. President 40 (Reagan) could claim to be the oldest.

UNIT 5D

QUICK QUIZ

1. **a**. Adding the data to a multiple bar graph simply requires a third bar for each year.
3. **a**. The color of Oregon corresponds to the $25,000 – $29,999 category.
5. **b**. The color within the small loop matches the band between contour lines marked by 40°F and 50°F, and thus the temperature there is between those values.
7. **c**. The height of the center of the pie charts corresponds to the number of women attending, and the height of the 1948 pie chart is about 450.
9. **c**. Each tick mark represents another power of ten.

DOES IT MAKE SENSE?

19. Does not make sense. Using three-dimensional bars for two-dimensional data is a cosmetic change; it adds no information.
21. Makes sense. Experimenting with different scales on the axes of graphs allows you to highlight certain aspects of the data.

BASIC SKILLS & CONCEPTS

23. a. China, India and Russia all had to import grain because their net grain production is negative.

b. China, India, and Russia are all expected to need grain imports because their projected 2030 net grain production is negative.

c. The graph shows that both China and India will require significant grain imports, and thus (based only on the information shown in the graph), it's likely world grain production will need to increase.

25. a. The death rate for pneumonia, cardiovascular disease, and tuberculosis decreased from 1900 to 2003; it increased for cancer.

b. The death rate was the greatest around 1950, where it was about 500 deaths per 100,000.

c. The death rate for cancer in 2000 was about 200 deaths per 100,000.

d. Deaths due to cardiovascular diseases will likely be responsible for the most deaths in 2050, as long as cancer levels out (as it has in the years following the mid-1990s).

27. a. About 15% of the budget went to net interest in 1990 and 1995; it dropped to about 8% in 2005.

b. In 1962, about 52% of the budget went to defense. The percentage dropped to about 23% in 1980, and dropped again in 2005, to about 20%.

c. In 1980, about 47% of the budget went to payments to individuals. This figure was about 57% in 2000, and about 54% in 2005.

29. There are significant regional differences: the probability that a black student had white classmates is much higher in the northern states. It's difficult to make a blanket statement about the differences between urban and rural areas, but it certainly appears that the probability is lower in many urban areas (look at San Francisco/Oakland, Los Angeles and San Diego, Chicago, Detroit, Miami, and the larger cities along the eastern seaboard, and compare these with adjoining rural areas).

31. The height of each bar corresponds to the percent of the population for a particular age group and for a given year. For example, in Figure 5.34a, we can see that in 1960, children less than 5 years in age made up about 10% of the population.

33. The percent of the youngest Americans has decreased.

35. The percentage of 45 to 65-year-olds was about 20% in 1960, and is projected to be about 27% in 2010.

37. The 45 to 65-year old group is projected to be largest in 2010.

39. The population is aging as evidenced by the increase in the over-65 category, and the decrease in the under-5 category.

43. In a flat pie chart, a wedge with central angle of, say, 60° represents 1/6 of the total area of the circle, no matter where it is located in the circle. In the distorted version shown, the face of the pie is an ellipse, and 60° wedges in ellipses don't necessarily have equal areas, and thus they won't correspond to the percentages they are attempting to depict.

45. Note that the zero point for braking distance is not on the graph, and because of this, it looks like the breaking distance of an Oldsmobile is about half that of a Lincoln. In reality, the breaking distance of Lincolns is about 11% greater than the breaking distance of Oldsmobiles. A fairer graph (not shown) would start from zero.

47. a. Public college costs rose by about 13% in the 2003-04 academic year.

b. The percentage increase in private college costs in the 2003-04 year was about 5%.

c. Private college costs rose more in dollars, because 5% of the '02–'03 costs for private colleges is more than 13% of the '02–'03 costs for public colleges.

FURTHER APPLICATIONS

49. Two graphs work best for this data. Both men and women are waiting much longer to get married for the first time.

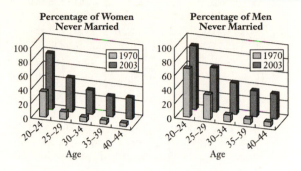

51. A simple multiple bar graph works well for this data.

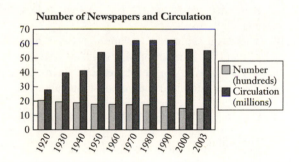

53. a. The change from one month to the next looks less significant when zero is included.

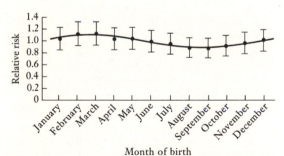

b. The claim is not justified because the error bars are large enough that the true values of relative risk may form a straight line across all months (which would mean the risk does not vary with season).

UNIT 5E

QUICK QUIZ

1. **c.** When one variable is correlated with another, an increase in one goes with either increasing values of the other, or decreasing values of the other.

3. **a.** When points on a scatter diagram all lie near a line, there is a strong correlation between the variables represented. We call it a positive correlation when an increase in one variable goes along with an increase in another (and this is what produces a positive slope).

5. **a.** Correlation is a necessary condition for causation.

7. **a.** When establishing causation, look for evidence that larger amounts of the suspected cause produce larger amounts of the effect.

9. **b.** From 1960 to 2000 (a span of 40 years), CO_2 concentration increased by about 50 ppm, and if that trend continues for another 40 years, to the year 2040, we can expect another increase of about 50 ppm, which will result in CO_2 levels near 420 ppm.

DOES IT MAKE SENSE?

17. Makes sense. A strong negative correlation means that a decrease in the price goes along with an increase in the number of tickets sold.

19. Does not make sense. Despite the fact that A and B are strongly correlated, we cannot be sure that A causes B (there may be an underlying effect that causes both A and B).

21. Makes sense. If it were true that an increase in E caused a decrease in F, then a negative correlation would exist.

BASIC SKILLS & CONCEPTS

23. a. We seek a correlation between a car's *weight* and its *city gas mileage*.
b. The diagram shows a moderately strong negative correlation.
c. Heavier cars tend to get lower gas mileage.

25. a. We seek a correlation between *salary level* and *percent of income given to charity* for employees of Big Co.
b. At first glance, there's no correlation (which is a reasonable answer), though one could also argue for a very weak negative correlation based on the light representation of data points in the upper right corner of the scatter diagram.
c. There's a slight indication that lower-salary employees give a larger percentage of their income to charity, but it's very weak.

27. Use degrees of latitude and degrees of temperature as units. There is a strong negative correlation, because as you move north from the equator, the temperatures in June tend to decrease.

29. Use years and hours as units. If the population studied consists of people from teen years to the elderly, one may find a weak negative correlation,

because teens probably spend more time on their cell phones than adults, particularly older adults.

31. Use people and dollars as units. There is no correlation because the average salary of a school teacher varies considerably from one region of the country to another, and this variation isn't in lock step with the population.

33. Use children per woman and years as units. There is a strong negative correlation, because an increase in the life expectancy of a country goes along with a decrease in the number of children born.

FURTHER APPLICATIONS

35. a.

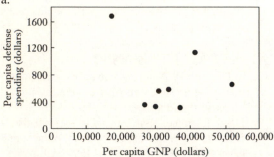

b. It appears there is a weak negative correlation.
c. In those countries where per capita GNP is high, there's more money available to spend on various government programs, and thus the per capita expenditure on defense decreases.

37. a.

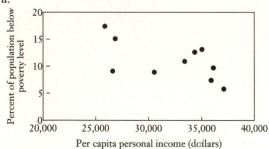

b. There is a weak negative correlation between the variables.
c. When the per capita income is higher, the average person is further away from poverty level, and the percent of the population below the poverty level decreases.

39. a.

b. There is a strong positive correlation.
c. The per pupil expenditure almost certainly includes teacher salaries (that is, salaries are part of the cost of educating the students), and thus in those districts where the teacher salaries are high, the per pupil expenditure is also high.

41. There is a positive correlation between the crime rate and the number of tourists. There may be direct cause and effect here: with more tourists, there are more targets for opportunistic crimes against them.

43. There is a negative correlation between the cost of gas and the number of SUVs sold. There may be a direct cause at work here, because when gas prices increase, people are more conscious about the gas mileage for the car they purchase.

45. Gas mileage is positively correlated with tire pressure. This is a direct cause: when tire pressure goes down, gas mileage also goes down.

47. a. When the suspected cause (bad air in the building in which you work) is removed, the afternoon headaches go away. This is guideline 2.
b. Even when you vary other factors, such as drinking Coke, the headaches are still present, suggesting that Coke isn't the cause, but rather bad air in the building is to blame. This is guideline 1.
c. When the suspected cause is decreased, the headaches are less frequent. This is guideline 3.
d. The observations suggest that your headaches are caused by bad air at your work environment.

49. People do not decide to become conductors until relatively late in life, say, 30 years of age or over. Having already survived this many years, a person has a longer-than-average life expectancy.

51. A plausible explanation is that many youth soccer leagues around the world use age divisions. Those children born earlier in a given year are slightly older than others born in the same year. Being slightly stronger and better developed, they survive and move on to more advanced leagues.

UNIT 6A

QUICK QUIZ

1. **b.** The mean of a data set is the sum of all the data values divided by the number of data points.
3. **c.** Outliers are data values that are much smaller or larger than the other values.
5. **c.** Because the mean is significantly higher than the median, the distribution is skewed to the right, which implies some students drink considerably more than 12 sodas per week. Specifically, it can be shown that at least one student must drink more than 16 sodas per week.
7. **a.** High outliers tend to pull the mean to the right of the median, and thus you would hope to get a salary near the mean. There are data sets with high outliers that can be constructed where the median is higher than the mean, but these are rare.
9. **c.** Driving in rush hour can go quickly on some days, but is agonizingly slow on others, so the time it takes to get downtown has high variation.

DOES IT MAKE SENSE?

17. Makes sense. The two highest grades may be large enough to balance the remaining seven lower grades so that the third highest is the mean.
19. Makes sense. When very high outliers are present, they tend to pull the mean to the right of the median (right-skewed).
21. Does not make sense. When the mean, median, and mode are all the same, it's a sign that you have a symmetric distribution of data.

BASIC SKILLS & CONCEPTS

23. The mean, median, and mode are 2.2698, 2.270, and 2.271, respectively. Because there is an even number of data points, the median is calculated by taking the mean of the middle two values (add 2.269 and 2.271, and divide by 2).
25. The mean, median, and mode are 0.215, 0.21, and 0.24, respectively.
27. The mean and median are 89.9 and 90, respectively. This is a bimodal data set, with modes of 87 and 91.
29. The mean and median are 0.81237 and 0.8161, respectively. The outlier is 0.7901, because it varies from the others by a couple hundredths of a pound, while the rest vary from each other only by a few thousandths of a pound. Without the outlier, the mean is 0.81608, and the median is 0.8163.
31. The median would be a better representation of the average. A small number of households with high earnings skew the distribution to the right, affecting the mean.
33. This distribution is probably skewed to the right by a small percentage of people who change jobs many times. Thus the median would do a better job of representing the average, as the outliers affect the mean.
35. This is a classic example of a symmetric distribution, and thus the mean and the median are almost identical (in a perfectly symmetric distribution, they are identical) – either the mean or the median could be used.
37. a. There is one peak on the far left due to all the people who scored low.
    b. The distribution is right-skewed because the scores tail off to the right.
    c. The variation is large (relative to the data in exercise 38, for example).
39. a. There are likely two peaks in such a distribution, one for the women, and one for the men.
    b. The distribution is symmetric because one would find a tail on the left representing the lightest people on the team, and a tail on the right representing the heaviest people on the team.
    c. Relative to a single-gendered set of athletes, the variation would be large.
41. a. Consider a typical data set for the monthly sales: we expect the sales to be high in the winter months, with very low sales (or even no sales) in the summer. The number of shovels sold each month, beginning in January, might look like this: {25, 20, 15, 3, 0, 0, 0, 0, 0, 3, 15, 20}. To picture this data set as a distribution, we create a frequency histogram.

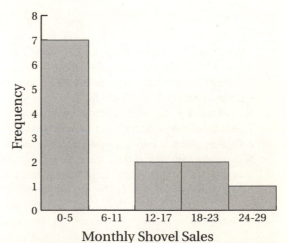

Monthly Shovel Sales

This histogram has two peaks, the second not as dramatic as the first. Realize that although the histogram would look different if we varied the bin sizes or the hypothetical sales data, its basic shape is captured in the diagram above. This is due to the fact that a majority of the monthly sales figures are expected to be low (corresponding to the left peak), and the winter monthly sales figures would be high (corresponding to the right peak).

b. The histogram shown in part **a** is right-skewed.

c. The variation is large because of the difference between low sales in the summer, and high sales in the winter.

43. a. There would be one peak in the distribution, corresponding to the average price of detergent.

b. More likely than not, there would be no extreme outliers in this data set, and it would be symmetric.

c. There would be moderate variation in the data set because we don't expect outliers. (The detergent may not sell well next to 19 other brands if its price is radically different. Of course this begs the question, "But what if it's much cheaper than the other brands?" That's not too likely to happen for a commodity like detergent, because once one company begins selling its detergent at a low price, the others would be forced to lower their prices to compete).

FURTHER APPLICATIONS

45. The distribution has two peaks (bimodal), with no symmetry, and large variation. For tourists who want to watch the geyser erupt, it means they may have to wait around for a while (most likely between 25 and 40 minutes, though perhaps as long as 50 to 90 minutes, depending on when they arrive at the site).

47. This distribution has one peak, is symmetric, and has moderate variation. The graph says that weights of rugby players are usually near the mean weight, but that there are some light and some heavy players.

49. a. We cannot be sure of the exact shape of the distribution with the given parameters, as we don't have all the raw data. However, because the mean is larger than the median, and because the data set has large outliers, it would likely be right-skewed, with a single peak at the mode.

b. About 150 families (50%) earned less than $35,000 because that value is the median income.

c. We don't have enough information to be certain how many families earned more than $41,000, but it is likely a little less than half, simply because the mean is a measure of center for all distributions. (Note that it is a little less than half because the outliers tend to pull the mean right of the median).

UNIT 6B

QUICK QUIZ

1. **a.** The range is defined as the high value – low value.

3. **a.** Roughly half of any data set is contained between the lower and upper quartiles ("roughly" because with an odd number of data values, one cannot break a data set into equal parts).

5. **b.** You need the high and low values to compute the range, and you need all of the values to compute the standard deviation, so the only thing you can compute is a single deviation.

7. **c.** The standard deviation is defined as the square root of the variance, and thus is always non-negative.

9. **b.** Newborn infants and first grade boys have similar heights, whereas there is considerable variation in the heights of all elementary school children.

DOES IT MAKE SENSE?

17. Does not make sense. The range depends upon only the low and high values, not on the middle values.

19. Makes sense. Consider the case where the 15 highest scores were 80, the next score was 68, and the lowest 14 scores were 40. There are numerous such cases that satisfy the conditions given.

21. Makes sense. The standard deviation describes the spread of the data, and one would certainly expect the heights of 5-year old children to have less variation than the heights of children aged 3 to 15.

BASIC SKILLS & CONCEPTS

23. The mean waiting time at Big Bank is (4.1 + 5.2 + 5.6 + 6.2 + 6.7 + 7.2 + 7.7 + 7.7 + 8.5 + 9.3 + 11.0) ÷ 11 = 79.2 ÷ 11 = 7.2. Because there are 11 values, the median is the sixth value = 7.2.

25. a. For the East Coast, the mean, median, and range are 134.97, 123.45, and 117.8, respectively. For the West Coast, the mean, median, and range are 145.82, 150.3, and 69.2, respectively.

b. For the East Coast, the five-number summary is (98.2, 108.7, 123.45, 140.0, 216.0). For the West Coast, it is (113.2, 122.7, 150.3, 156.0, 182.4). The boxplots are not shown.

c. The standard deviation for the East Coast is 42.86. For the West Coast, it is 25.06.

d. For the East Coast, the standard deviation is approximately 117.8 ÷ 4 = 29.45, which is a far cry from the actual value of 42.86. This is due, in part, to the outlier of New York City (216.0). For the West Coast, the standard deviation is approximately 69.2 ÷ 4 = 17.3, which is also well off the mark.

e. The cost of living index is smaller, on average, for the East Coast cities shown, though the variation is higher (due in part to the outlier of New York City).

27. a. In San Francisco, the mean, median, and range are 65.1, 65, and 13, respectively. In Chicago, the mean, median, and range are 58.3, 60, and 54, respectively.

b. The five-number summary for San Francisco is (58, 62, 65, 68.5, 71). It is (30, 40.5, 60, 76.5, 84) in Chicago. The boxplots are not shown.

c. The standard deviation in San Francisco is 4.3; in Chicago, it is 19.6.

d. The standard deviation in San Francisco is approximately 13 ÷ 4 = 3.25, which underestimates the actual value of 4.3. In Chicago, the standard deviation is approximately 54 ÷ 4 = 13.5, which is considerably lower than the actual value of 19.6.

e. The average high temperatures in San Francisco are a little warmer than those in Chicago, though the variation of temperatures is much smaller in San Francisco.

FURTHER APPLICATIONS

29. a.

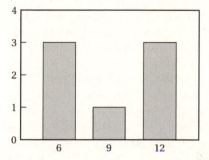

b. The five-number summaries for each of the sets shown are (in order): (9, 9, 9, 9, 9); (8, 8, 9, 10, 10); (8, 8, 9, 10, 10); and (6, 6, 9, 12, 12). The boxplots are not shown.

c. The standard deviations for the sets are (in order): 0.000, 0.816, 1.000, and 3.000.

d. Looking at just the answers to part **c**, we can tell that the variation gets larger from set to set.

31. The first pizza shop has a slightly larger mean delivery time, but much less variation in delivery time when compared to the second shop. If you want a reliable delivery time, choose the first shop. If you don't care when the pizza arrives, choose the shop that offers cheaper pizza (you like them equally well, so you may as well save some money).

33. A lower standard deviation means more certainty in the return, and a lower risk.

35. Batting averages are less varied today than they were in the past. Because the mean is unchanged, batting averages above .350 are less common today.

UNIT 6C

QUICK QUIZ

1. **b**. A normal distribution is symmetric with one peak, and this produces a bell-shaped curve.

3. **a**. Data values farther from the mean correspond to lower frequencies than those close to the mean, due to the bell-shaped distribution.

5. **a**. Roughly 68% of data values fall within one standard deviation of the mean, and 68% is about 2/3.

7. **a**. Note that 43 mpg is 1 standard deviation above the mean. Since 68% of the data lies within one standard deviation of the mean, the remaining 32% is farther than 1 standard deviation from the mean. The normal distribution is symmetric, so half of 32%, or 16%, lies above 1 standard deviation.

9. **c**. If your friend said his IQ was in the 75th percentile, it would mean his IQ is larger than 75% of the population. It is impossible to have an IQ that is larger than 102% of the population.

DOES IT MAKE SENSE?

15. Makes sense. Physical characteristics are often normally distributed, and college basketball players are tall, though not all of them, so it makes sense that the mean is 6'3", with a standard deviation of 3" (this would put approximately 95% of the players within 5'9" to 6'9").

17. Does not make sense. With a mean of 6.8 pounds, and a standard deviation of 7 pounds, you would expect that a small percentage of babies would be born 2 standard deviations above the mean. But this would imply that some babies weigh more than 20.8 pounds, which is much too heavy for a newborn.

19. Makes sense. A standard score of 2 or more corresponds to percentiles of 97.72% and higher, so while this could certainly happen, the teacher is giving out As (on the final exam) to only 2.28% of the class

BASIC SKILLS & CONCEPTS

21. Diagrams **a** and **c** are normal, and diagram **c** has a larger standard deviation because its values are more spread out.

23. Most trains would probably leave near to their scheduled departure times, but a few could be considerably delayed, which would result in a right-skewed (and non-normal) distribution.

25. Random factors would be responsible for the variation seen in the distances from the bull's-eye, so this distribution would be normal.

27. Exams such as the SAT are designed to produce scores that are normally distributed, and most difficult exams have scores that are similarly normally distributed.

29. a. Half of any normally distributed data set lies below the mean, so 50% of the scores are less than 100.

b. Note that 120 is one standard deviation above the mean. Since 68% of the data lies within one

standard deviation of the mean (between 80 and 120), half of 68%, or 34%, lies between 100 and 120. Add 34% to 50% (the answer to part **a**) to find that 84% of the scores are less than 120.

c. Note that 140 is two standard deviations above the mean. Since 95% of the data lies within two standard deviations above the mean (between 60 and 140), half of 95%, or 47.5%, lies between 100 and 140. Add 47.5% to 50% (see part **a**) to find that 97.5% of the scores are below 140.

d. Note that 60 is two standard deviations below the mean. Since 95% of the data lies within two standard deviations of the mean (between 60 and 140), 5% of the data lies outside this range. Because the normal distribution is symmetric, half of 5%, or 2.5%, lies in the lower tail below 60.

e. A sketch of the normal distribution would show that the amount of data lying above 60 is equivalent to the amount of data lying below 140, so the answer is 97.5% (see part **c**).

f. Note that 160 is three standard deviations above the mean. Since 99.7% of the data lies within three standard deviations of the mean (between 40 and 160), 0.3% of the data lies outside this range. Because the normal distribution is symmetric, half of 0.3%, or 0.15%, lies in the upper tail above 160.

g. A sketch of the normal distribution shows that the percent of data lying above 80 is equivalent to the percent lying below 120, so the answer is 84% (see part **b**).

h. The percent of data between 60 and 140 is 95%, because these are each two standard deviations from the mean.

31. A score of 75 is one standard deviation above the mean. Since 68% of normally distributed data lies within one standard deviation of the mean, half of 68%, or 34%, lies between 67 and 75 due to symmetry. The percent of data lying below 67 is 50%, so we add these two results to find that 84% of the scores lie below 75.

33. Subtract two standard deviations from the mean to get the cutoff score of 67 − 16 = 51. Since 95% of the data lies within two standard deviations of the mean, 5% lies outside this range. Half of 5%, or 2.5%, will be below 51, and will fail the exam.

35. The standard score for 67 is $z = (67 − 67)/8 = 0$.

37. The standard score for 55 is $z = (55 − 67)/8 = −1.5$

39. a. The data value is one standard deviation below the mean, so $z = −1$. The corresponding percentile is 15.87%.

b. $z = 1.5$, and the corresponding percentile is 93.32%.

c. $z = 2$, and the corresponding percentile is 97.72%.

41. a. The 15th percentile corresponds to a standard score of about $z = -1$, which means the data value is one standard deviation below the mean.
b. The 75th percentile corresponds to a standard score of about $z = 0.70$, which means the data value is 0.7 standard deviations above the mean.
c. The 23rd percentile corresponds to a standard score of about $z = -0.75$, which means the data value is 0.75 standard deviations below the mean.

FURTHER APPLICATIONS

43. About 68% of births occur within 15 days of the mean, because this range is one standard deviation from the mean.

45. Since 68% of births occur within 15 days of the due date, 32% occur outside this range, and half of 32%, or 16%, occur more than 15 days before the due date. It's probably a bad idea to travel that close to one's due date, because there's about a 1/6 chance of going into labor.

47. a. The standard score is $z = (66 - 63.6)/2.5 = 0.96$, which corresponds to an approximate percentile of 83%.
b. $z = (62 - 63.6)/2.5 = -0.64$, which corresponds to an approximate percentile of 26%.
c. $z = (61.5 - 63.6)/2.5 = -0.84$, which corresponds to an approximate percentile of 20%.
d. $z = (64.7 - 63.6)/2.5 = 0.44$, which corresponds to an approximate percentile of 67%.

49. It is not likely because heights of eighth-graders would be normally distributed, and a standard deviation of 40 inches would imply some of the eighth graders (about 16%) are taller than 95 inches, which is preposterous.

51. The 90th percentile corresponds to a z-score of about 1.3, which means a data value is 1.3 standard deviations above the mean. For the verbal portion of the GRE, this translates into a score of $469 + 1.3(120) = 625$. For the quantitative portion, it translates into a score of $597 + 1.3(148) = 789$.

53. The standard score for 600 is $z = (600 - 597)/148 = 0.02$, which corresponds to about the 51st percentile. This is the percent of students who score below 600, so the percent that score above 600 is 49%.

55. The z-score for an 800 on the quantitative exam is $z = (800 - 597)/148 = 1.37$, or about 1.4. This corresponds to the 91.92 percentile, so about 92% score below 800, which means 8% score 800.

57. For the verbal exam, $z = (400 - 469)/120 = -0.575$. This is exactly between two entries on Table 6.3: $z = -0.60$, and $z = -0.55$. The percentiles corresponding to these z-scores are 27.43% and 29.12% (respectively), so about 28% score below 400. For the quantitative exam, $z = (400 - 597)/148 = -1.33$, which corresponds to an

approximate percentile of 9%. Thus about 9% score below 400.

UNIT 6D

QUICK QUIZ

1. **c.** A divorce rate of 60% is statistically significant because such large variation from the national rate of 30% is not likely to occur by chance alone.

3. **a.** Statistical significance at the 0.01 level means an observed difference between the effectiveness of the remedy and the placebo would occur by chance in fewer than 1 out of 100 trials of the experiment.

5. **b.** The confidence interval is found by subtracting and adding 3% to the results of the poll (65%).

7. **c.** The margin of error is approximately $\frac{1}{\sqrt{n}}$. Set this equal to 4% (0.04), and solve for n to find that the sample size was 625. If we quadruple the sample size to 2500, the margin of error will be approximately $\frac{1}{\sqrt{2500}} = 0.02$, or 2%.

9. **c.** If you cannot reject the null hypothesis, the only thing you can be sure of is that you don't have evidence to support the alternative hypothesis.

DOES IT MAKE SENSE?

19. Makes sense. If the number of people cured by the new drug is not significantly larger than the number cured by the old drug, the difference could be explained by chance alone, which means the results are not statistically significant.

21. Does not make sense. The margin of error is based upon the number of people surveyed (the sample size), and thus both Agency A and B should have the same margin of error.

23. Makes sense. The alternative hypothesis is the claim that is accepted if the null hypothesis is rejected.

BASIC SKILLS & CONCEPTS

25. Most of the time you would expect around 25 tails in 50 tosses of a coin. It would be very rare to get only 3 tails, and thus this is statistically significant.

27. Note that 11/200 is about 5% – this is exactly what you would expect by chance alone from an airline that has 95% of its flights on time.

29. There aren't many winners in the Power Ball lottery, and yet there are a good number of 7-11 stores. It would be very unlikely that ten winners

in a row purchased their tickets at the same store, so this is statistically significant.

31. The result is significant at far less than the 0.01 level (which means it is significant at both the 0.01 and 0.05 level), because the probability of finding the stated results is less than 1 in 1 million by chance alone. This gives us good evidence to support the alternative hypothesis that the accepted value for temperature is wrong.

33. The results of the study are significant at neither the 0.05 level nor the 0.01 level, and thus the improvement is not statistically significant.

35. The margin of error is $1/\sqrt{1012} = 0.03$, or 3%, and the 95% confidence interval is 29% to 35%. This means we can be 95% confident that the actual percentage is between 29% and 35%.

37. The margin of error is $1/\sqrt{1002} = 0.03$, or 3%, and the 95% confidence interval is 27% to 33% for the satisfied group, and 64% to 70% for the dissatisfied group.

39. The margin of error is $1/\sqrt{1003} = 0.03$, or 3%, and the 95% confidence interval is 45% to 51%.

41. The margin of error is $1/\sqrt{2435} = 0.02$, or 2%, and the 95% confidence intervals are 15% to 19% and 12% to 16%.

43. a. Null hypothesis: percentage of high school graduates = 85%. Alternative hypothesis: percentage of high school students > 85%.
b. Rejecting the null hypothesis means there is evidence that the percentage of high school graduates exceeds 85%. Failing to reject the null hypothesis means there is insufficient evidence to conclude that the percentage of high school graduates exceeds 85%.

45. a. Null hypothesis: mean teacher salary = $47,750. Alternative hypothesis: mean teacher salary > $47,750.
b. Rejecting the null hypothesis means there is evidence that the mean teacher salary in the state exceeds $47,750. Failing to reject the null hypothesis means there is insufficient evidence to conclude that the mean teacher salary exceeds $47,750.

47. a. Null hypothesis: percentage of underrepresented students = 20%. Alternative hypothesis: percentage of underrepresented students < 20%.
b. Rejecting the null hypothesis means there is evidence that the percentage of underrepresented students is less than 20%. Failing to reject the null hypothesis means there is insufficient evidence to conclude that the percentage of underrepresented students is less than 20%.

49. Null hypothesis: mean annual mileage of cars in the fleet = 11,725 miles. Alternative hypothesis: mean annual mileage of cars in the fleet > 11,725.

The result is significant at the 0.01 level, and provides good evidence for rejecting the null hypothesis.

51. Null hypothesis: mean stay = 2.1 days. Alternative hypothesis: mean stay > 2.1 days. The result is not significant at the 0.05 level, and there are no grounds for rejecting the null hypothesis.

53. Null hypothesis: mean income = $40,000. Alternative hypothesis: mean income > $40,000. The result is significant at the 0.01 level, and provides good evidence for rejecting the null hypothesis.

FURTHER APPLICATIONS

55. The margin of error is $1/\sqrt{13,000} = 0.009$, or 0.9%, and the 95% confidence interval is 64.1% to 65.9%. This means we can be 95% confident that the actual percentage of viewers who watched *American Idol* was between 64.1% and 65.9%.

57. The margin of error is approximately $1/\sqrt{n}$. If we want to decrease the margin of error by a factor of 2, we must increase the sample size by a factor of 4, because $\frac{1}{\sqrt{4n}} = \frac{1}{2\sqrt{n}} = \frac{1}{2} \cdot \frac{1}{\sqrt{n}}$.

59. The margin of error, 3%, is consistent with the sample size because $1/\sqrt{1019} = 0.03 = 3\%$.

UNIT 7A

QUICK QUIZ

1. **b**. HHT is a different outcome than HTH due to the order in which the heads appear, but both are the same event because each has two heads.

3. **a**. This is a theoretical probability because we are assuming each outcome is equally likely, and no experiment is being carried out to determine the probability.

5. **b**. Table 7.3 shows that the probability Mackenzie wins is 5/36, whereas the probability that Serena wins is 4/36.

7. **b**. The lowest sum that could occur when four dice are rolled is 4, and the highest is 24, and every sum in between is possible.

9. **a**. If Triple Treat's probability of winning is 1/4, then the probability he loses is 3/4. This implies the odds that he wins is (1/4) ÷ (3/4), or 1 to 3, which, in turn, means the odds against are 3 to 1.

DOES IT MAKE SENSE?

17. Makes sense. The outcomes are HTTT, THTT, TTHT, TTTH.

19. Makes sense. This is a subjective probability, but it's as good as any other guess one might make.

21. Does not make sense. The sum of $P(A)$ and $P($not $A)$ is always 1.

BASIC SKILLS & CONCEPTS

23. There are $12 \times 3 = 36$ choices.

25. There are $4 \times 6 \times 10 \times 12 = 2880$ systems.

27. There are $28 \times 4 = 112$ combinations.

29. There are $10 \times 5 \times 8 = 400$ systems.

31. There are 4 kings in a standard deck, so the probability is 4/52 = 1/13. We are assuming that each card is equally likely to be chosen.

33. The probability is 2/6 = 1/3. We are assuming equally likely outcomes.

35. There are 12 royalty cards, so the probability is 12/52 = 3/13. It is assumed that each card is equally likely to be chosen.

37. The probability is 2/52 = 1/26 because there are 2 black jacks. We are assuming equally likely outcomes.

39. There are ten possible outcomes for the last digit in a phone number, and one of these is a 5, so the probability is 1/10. It is assumed that each digit is equally likely to occur at the end of a phone number.

41. There are 4 possible outcomes (BB, BG, GB, GG), and one of these consists of two boys, so the probability is 1/4. This assumes each outcome is equally likely.

43. There are 8 possible outcomes with a 3-child family (BBB, BBG, BGB, GBB, GGB, GBG, BGG, GGG), and 3 of these have exactly one boy, so the probability is 3/8. This assumes each outcome is equally likely to occur.

45. There are 22 pairs of socks, and 8 are white, so the probability is 8/22 = 4/11, assuming that each pair of socks is equally likely to be chosen.

47. Your address ends in one of ten digits, and the probability that a randomly selected person has the same ending digit is 1/10. This assumes that each digit is as likely as the others to be at the end of an address, and it assumes that addresses end in digits.

49. The probability is 18/30 = 3/5 = 0.6.

51. The probability is 76% = 0.6.

53. The probability is 225/250 = 0.9.

55. Answers will vary.

57. The probability is 5/6, because there are 5 outcomes that aren't 6s. It is assumed that each outcome is equally likely.

59. The probability he will make the next free throw is 86% = 0.86, so the probability he will miss is 1 − 0.86 = 0.14, or 14%. The probabilities computed here are based on his past performance.

61. The probability of rolling a sum of 7 is 6/36 (see Table 7.3), so the probability of not rolling a sum of 7 is 1 − 6/36 = 30/36 = 5/6. Equally likely outcomes are assumed.

63. The probability of having a 100-year flood is 1/100, so the probability of not having one is 1 − 1/100 = 99/100 = 0.99. This assumes that each year the probability of a 100-year flood is 1/100.

65.

Result	Probability
0 H	1/16
1 H	4/16
2 H	6/16
3 H	4/16
4 H	1/16

67. The probability of getting a 1 or a 2 is 2/6 = 1/3, and the probability of not getting a 1 or a 2 is 2/3. Thus the odds for the event are (1/3) ÷ (2/3) = 1 to 2, and the odds against are 2 to 1.

69. This problem is similar to exercise 67, except that we are focusing on getting a 5 or a 6 (instead of a 1 or a 2) – the probabilities and odds remain the same. Thus the odds for the event are 1 to 2, and odds against are 2 to 1.

71. Since the odds on your bet are 3 to 4, you'll win $3 for every $4 you bet. A $20 bet is five $4 bets, so you will gain $15.

FURTHER APPLICATIONS

73. a. There are 52 women at the convention, so the probability a random delegate is a woman is 52/100 = 0.52.

b. There are 10 Independents, so the probability is 10/100 = 0.1.

c. There are 41 Democrats, and thus 59 non-Democrats, which means the probability is 59/100 = 0.59.

d. There are 21 female Republicans, so the probability is 21/100 = 0.21.

e. There are 16 male Democrats, so the probability you meet one is 16/100 = 0.16. Thus the probability that you meet a someone who is not a male Democrat is 1 − 0.16 = 0.84.

75. The empirical probabilities for 0 heads, 1 head, 2 heads, and 3 heads are (respectively): 260/1000 = 0.26, 495/1000 = 0.495, 245/1000 = 0.245, and 0/1000 = 0.0. The 8 possible outcomes when 3 coins are tossed are (HHH, HHT, HTH, THH, TTH, THT, HTT, TTT), and thus the theoretical probabilities for 0 heads, 1 head, 2 heads, and 3 heads are (respectively): 1/8 = 0.125, 3/8 = 0.375, 3/8 = 0.375, and 1/8 = 0.125. The differences between the empirical and theoretical probabilities are large enough to suspect that the coins are unfair.

77. The odds for event A are $(0.99) \div (0.01) = 99$ to 1, and the odds for event B are $(0.96) \div (0.04) = 24$ to 1. The odds for A are much higher, even though the probabilities for A and B are relatively close to one another.

UNIT 7B

QUICK QUIZ

1. **c.** The probability of getting a double-2 on at least one of the rolls is $1 - (35/36)^2$.

3. **b.** The probability of choosing a pink rose on the second pick is dependent upon what happens on the first pick (because there will be fewer roses in the vases).

5. **c.** A person can be born on both a Wednesday and in July, so the events are overlapping.

7. **a.** Since these events are independent, we have $P(sum\ of\ 3\ and\ sum\ of\ 4) = P(sum\ of\ 3) \times P(sum\ of\ 4) = 2/36 \times 3/36$.

9. **c.** This answer shows the correct application of the *at least once* rule.

DOES IT MAKE SENSE?

15. Makes sense. You can't get both heads and tails on a single flip of a coin, but you must get one or the other.

17. Does not make sense. The probability of drawing an ace or a spade is 16/52, but the probability of drawing the ace of spades is 1/52.

19. Does not make sense. The chance of getting at least one 5 is $1 - (5/6)^3 \neq 3/6$.

BASIC SKILLS & CONCEPTS

21. The events are independent. $P(6\ and\ 6\ and\ 1\ and\ 1) = P(6) \times P(6) \times P(1) \times P(1) = (1/6)^4 = 1/1296$.

23. The events are dependent. $P(A_1\ and\ A_2\ and\ A_3) = P(A_1) \times P(A_2\ given\ A_1) \times P(A_3\ given\ A_1\ and\ A_2) = (4/52) \times (3/51) \times (2/50) = 0.00018$.

25. The events are independent. Because the probability of rolling a 6 is the same on each throw (1/6), the probability of getting all 6s is $(1/6)^3 = 1/216 = 0.0046$.

27. The events are dependent. This is similar to exercise 23, with an answer of $(13/52) \times (12/51) \times (11/50) = 0.0129$.

29. The events are most likely independent (there may be some situations where one could argue dependence, such as the case where a woman gives birth to quintuplets, and these happen to be the next five births – but we can't solve the problem without assuming independence). Assuming the probability of giving birth to a girl is 1/2, the answer is $(1/2)^5 = 1/32$.

31. The events are dependent. The probability is $(18/30) \times (17/29) \times (16/28) \times (15/27) = 0.1117$.

33. The events are independent. The probability is $(1/12)^3 = 1/1728 = 0.00058$.

35. The events are independent for all practical purposes (as long as the student population is large). The probability is $(1/12)^3 = 1/1728$.

37. The events are non-overlapping. $P(red\ 6\ or\ black\ 8) = P(red\ 6) + P(black\ 8) = 2/52 + 2/52 = 1/13$.

39. The events are non-overlapping. $P(2\ or\ 3\ or\ 4) = P(2) + P(3) + P(4) = 1/36 + 2/36 + 3/36 = 1/6$.

41. The events are non-overlapping. $P(Q\ or\ K) = P(Q) + P(K) = 4/52 + 4/52 = 2/13$.

43. The events are overlapping. The probability is $1/6 + 3/6 - 1/6 = 3/6 = 1/2$.

45. The events are overlapping. $P(W\ or\ R) = P(W) + P(R) - P(W\ and\ R) = 70/100 + 60/100 - 40/100 = 90/100 = 0.9$.

47. The events are non-overlapping. The easiest way to solve the problem is to look at the set of possible outcomes for a three child family, (GGG, GGB, GBG, BGG, BBG, BGB, GBB, BBB), and

count the number of outcomes with either 1 or 2 girls. The probability is 6/8 = 3/4. One could also use the formulas for *or* probabilities.

49. The probability of not getting a 3 on a single roll is 5/6, so P(at least one 3) $= 1 - (5/6)^4 = 0.518$.

51. P(at least one 2) $= 1 - (5/6)^3 = 0.421$.

53. P(at least one ace) $= 1 - (48/52)^{10} = 0.551$.

55. P(rain at least once) $= 1 - (0.80)^5 = 0.672$.

57. P(at least one flu encounter) $= 1 - (0.96)^{10} = 0.335$.

59. P(at least one ticket) $= 1 - (0.7)^5 = 0.832$.

FURTHER APPLICATIONS

61. a. $P(G$ or $P) = P(G) + P(P) - P(G$ and $P) = 956/1028 + 450/1028 - 392/1028 = 0.986$.
    b. $P(NG$ or $NP) = P(NG) + P(NP) - P(NG$ and $NP) = 72/1028 + 578/1028 - 14/1028 = 0.619$.

63. a. These events are dependent because there are fewer people to choose from on the second and subsequent calls, and this affects the probability that a Republican is chosen.
    b. $P(R_1$ and $R_2) = P(R_1) \times P(R_2$ given $R_1) = (25/45) \times (24/44) = 0.303$.
    c. If the events are treated as independent, then the pollster is selecting a person from random and not crossing that person off the list of candidates for the second call. The probability of selecting two Republicans is then $P(R_1$ and $R_2) = P(R_1) \times P(R_2) = (25/45) \times (25/45) = 0.309$.
    d. The probabilities are nearly the same. This is due to the fact that the list of names is large, and removing one person from the list before making the second call doesn't appreciably change the probability that a Republican will be chosen.

65. The probability of winning would be $1 - (35/36)^{25} = 0.506$. Since this is larger than 50%, he should come out ahead over time.

67. a. The probability is 1/20.
    b. $P(H_1$ and $H_2) = P(H_1) \times P(H_2) = (1/20) \times (1/20) = 1/400 = 0.0025$.
    c. P(at least one hurricane) $= 1 - (19/20)^{10} = 0.401$.

UNIT 7C

QUICK QUIZ

1. **b.** The law of large numbers says that the proportion of years should be close to the probability of a hurricane on a single year (and $100/1000 = 0.1$).

3. **c.** If you purchase 1/10 of the available number of lottery tickets, there's a 1 in 10 chance that you have the winning ticket. But this means that 90% of the time you have purchased 1 million losing tickets.

5. **b.** The expected value of a $2000 fire insurance policy from the point of view of the insurance company is $\$2000 \times 1 + (-\$100,000) \times 1/50 = 0$, so they can expect to lose money if they charge less than $2000 per policy.

7. **c.** The results of the previous five games have no bearing on the outcome of the next game.

9. **c.** In order to compute the probability of winning on any single trial, you would need to know the probability of winning on each of the various payouts.

DOES IT MAKE SENSE?

17. Makes sense. An organization holding a raffle usually wants to raise money, which means the expected value for purchasers of raffle tickets will be negative.

19. Does not make sense. Each of the 16 possible outcomes when four coins are tossed is equally likely.

21. Does not make sense. The slot machine doesn't remember what happened on the previous 25 pulls, and your probability of winning remains the same for each pull.

BASIC SKILLS & CONCEPTS

23. You should not expect to get exactly 5000 heads (the probability of that happening is quite small), but you should expect to get a result near to 5000 heads. The law of large numbers says that the proportion of a large number of coin tosses that result in heads is near to the probability of getting heads on a single toss. Since the probability of getting heads is 1/2, you can expect that around one-half of the 10,000 tosses will result in heads.

25. The probability of tossing three heads is 1/8, and that of not tossing three heads is 7/8. Thus the expected value of the game is $\$6 \times 1/8 + (-\$1) \times 7/8 = -\$0.125$. Though one can assign a probability to the outcome of a single game, the result cannot be predicted, so you can't be sure whether you will win or lose money in one game. However, over the course of 100 games, the law of averages comes into play, and you can expect to lose money.

27. The probability of rolling two even numbers is $9/36 = 1/4$ (of the 36 possible outcomes, 9 are "two even numbers"), and the probability of not rolling two even numbers is $27/36 = 3/4$. Thus the expected value is $\$2 \times 1/4 + (-\$1) \times 3/4 = -\$0.25$. The outcome of one game cannot be predicted, though over 100 games, you should expect to lose.

29. The expected value for a single policy is $800 × 1 + (–$10,000) × 1/50 + (–$20,000) × 1/100 + (–$50,000) × 1/250 = $200. The company can expect a profit of $200 × 10,000 = $2 million.

31. For the sake of simplicity, assume that you arrive on the half-minute for each minute of the hour (that is, you arrive 30 seconds into the first minute of the hour, 30 seconds into the second minute of the hour, etc.). The probability that you arrive in any given minute in the first half-hour is 1/30, and your wait times (in minutes) for the first 30 minutes are 29.5, 28.5, 27.5, … , 0.5. Thus your expected wait time is 29.5 × 1/30 + 28.5 × 1/30 + 27.5 × 1/30 + … + 0.5 × 1/30 = 15 minutes. Of course the same is true for the second half-hour. (Note for instructors: if you wanted to allow for arrival at *any* time during a given minute, you would need to use calculus in order to solve the problem with any level of rigor, because the wait time function is continuous).

33. a. Heads came up 46% of the time, and you've lost $8.
    b. Yes, the increase in heads is consistent with the law of large numbers: the more often the coin is tossed, the more likely the percentage of heads is closer to 50%. Since 47% of 300 is 141, there have been 141 heads and 159 tails, and you are now behind $18.
    c. You would need 59 heads in the next 100 tosses (so that the total number of heads would be 200) in order to break even. While it's possible to get 59 heads in 100 tosses, a number nearer to 50 is more likely.
    d. If you decide to keep playing, the most likely outcome is that half of the time you'll see heads, and half the time you'll see tails, which means you'll still be behind.

35. a. The probability of getting heads is the same as the probability of getting tails.
    b. The most likely outcome after 1000 more flips is 500 heads and 500 tails (in which case the difference would remain at 16). The distribution of the number of heads is symmetric so that any deviation from "500 heads/500 tails" is as likely to be in the upper tail as the lower tail.
    c. Once you have fewer heads than tails, the most likely occurrence after additional coin tosses is that the deficit of heads remains.

37. The probability of getting ten heads in ten tosses is $(1/2)^{10} = 1/1024$. There are many ten-toss sequences in the 1000 tosses (991 to be precise), and you should expect, on average, one streak of ten heads in every 1024 ten-toss sequences.

39. a. Suppose you bet $1. Then your expected value is $1 × 0.493 + (–$1) × 0.507 = –$0.014. In general, the expected value is –1.4% of your bet,

which means the house edge is 1.4% (or 1.4¢ per dollar bet).
    b. You should expect to lose 100 × $0.014 = $1.40.
    c. You should expect to lose $500 × 1.4% = $7.
    d. The casino should expect to earn 1.4% of $1 million, which is $14,000.

FURTHER APPLICATIONS

41. The expected value is (–$1) × 1 + $30,000,000 × 1/80,089,128 + $100,000 × 1/1,953,393 + $5000 × 1/364,042 + $100 × 1/8466 + $7 × 1/605 + $4 × 1/188 + $3 × 1/74 = –$0.48. You can expect to lose 365 × $0.48 = $175.20 (or, if you don't round the intermediate answer of expected value, you can expect to lose around $173.45).

43. The expected value is (–$1) × 1 + $5,000,000 × 1/76,275,360 + $150,000 × 1/2,179,296 + $5000 × 1/339,002 + $150 × 1/9686 + $100 × 1/7705 + $5 × 1/220 + $2 × 1/102 + $1 × 1/62 = –$0.76. You can expect to lose 365 × $0.76 = $277.40 (or, if you don't round the intermediate answer of expected value, you can expect to lose around $278.84).

45. The expected value for 1-point kicks is 1 × 0.94 + 0 × 0.06 = 0.94 points. The expected value for 2-point attempts is 2 × 0.37 + 0 × 0.63 = 0.74 points. In most cases, it makes sense to take the almost certain extra-point, rather than risk a failed 2-point attempt. One situation where most coaches will go for a 2-point attempt: if your team is down by 8 points late in the game, and you score a touchdown (6 points), you don't benefit much by taking the extra point. You'll still need a field goal (3 points) to win the game. However, if you go for a 2-point attempt, you may tie the game, and send it into overtime. If you miss the 2-point attempt, you'll still be able to win the game with a field goal.

47. The expected value for the number of people in a household is 1.5 × 0.57 + 3.5 × 0.32 + 6 × 0.11 = 2.6 people. The expected value is a weighted mean, and it is a good approximation to the actual mean as long as the midpoints are representative of the categories.

49. a. For decision 1, the expected value for option A is $1,000,000, and the expected value for option B is $2,500,000 × 0.1 + $1,000,000 × 0.89 + $0 × 0.01 = $1,140,000. For decision 2, the expected value for option A is $1,000,000 × 0.11 + $0 × 0.89 = $110,000, and the expected value for option B is $2,500,000 × 0.1 + $0 × 0.9 = $250,000.
    b. The responses in decision 1 are not consistent with the expected values, whereas they are for decision 2.

c. For decision 1, people aren't willing to risk it all and earn nothing, and thus they chose option A – there is a huge guaranteed payback (especially for people in the 1950s) that isn't much different from the expected value of option B. For decision 2, people realize that there's little chance for gaining anything, so they may as well go for the bigger prize at the expense of a slightly lowered chance of winning it.

UNIT 7D

QUICK QUIZ

1. **a.** The accident rate will go down because the numerator of the "accident rate per mile" ratio is decreasing while the denominator is increasing.

3. **b.** Though there has been some variation in the number of deaths, the number of miles driven has steadily increased, and thus the death rate has generally decreased.

5. **c.** The death rate per person is $\frac{105,695}{300 \text{ million}}$, and this needs to be multiplied by 100,000 to get the death rate per 100,000 people.

7. **c.** Your life expectancy decreases as you age because the number of years before you die is decreasing.

9. **b.** The life expectancies of men are consistently below those of women (and from the graph, the gap has widened a bit in the last several decades).

DOES IT MAKE SENSE?

15. Does not make sense. Automobiles are successfully sold in huge quantities despite the risks involved.

17. Does not make sense. Your life expectancy is an average life expectancy for people born when you were. Much more important is your genetic makeup, and the choices you make throughout life (your diet, your line of work, the activities you choose to participate in, etc.).

BASIC SKILLS & CONCEPTS

19. The per-mile death rate was $\frac{42,646}{2.9 \times 10^{12}} = 1.5 \times 10^{-8}$.

    Multiply by 100 million to get 1.5 deaths per 100 million vehicle-miles.

21. The per-driver rate was $\frac{42,646}{1.98 \times 10^{8}} = 2.15 \times 10^{-4}$.

    Multiply by 100,000 to get 21.5 fatalities per 100,000 drivers.

23. There were $\frac{22}{1.87 \times 10^{7}} \times 10^{5} = 0.12$ death per 100,000 flight hours, $\frac{22}{7.7 \times 10^{9}} \times 10^{9} = 2.9$ deaths per 1 billion miles flown, and $\frac{22}{1.07 \times 10^{7}} \times 10^{5} = 0.21$ death per 100,000 departures. There were $\frac{32}{1.87 \times 10^{7}} \times 10^{5} = 0.17$ accident per 100,000 flight hours, $\frac{32}{7.7 \times 10^{9}} \times 10^{9} = 4.2$ accidents per 1 billion miles flown, and $\frac{32}{1.07 \times 10^{7}} \times 10^{5} = 0.30$ accident per 100,000 departures.

25. The probability of death by diabetes is 73,249 ÷ 300 million = 0.00024. Since 73,249 ÷ 34,243 = 2.1, death by diabetes is 2.1 times more likely than death by septicemia.

27. The death rate is $\frac{63,343}{3 \times 10^{8}} \times 10^{5} = 21.1$ deaths per 100,000 people.

29. The death rate due to stroke is $\frac{157,803}{3 \times 10^{8}} \times 10^{5} = 52.6$ deaths per 100,000 people, so in a city of 500,000, you would expect 52.6 × 5 = 263 people to die from a stroke each year.

31. The death rate for 60-year-olds is about 13 per 1000, and for 65-year-olds, it's about 20 per 1000, so the death rate for the category of 60- to 65-year-olds is about 16 deaths per 1000 people. Multiply 16/1000 by 12.6 million to find the number of people in this category expected to die in a year: $\frac{16}{1000} \times 12.6 \times 10^{6} = 201,600$ people.

33. The life expectancy for a 40-year-old is about 40 years, so the average 40-year-old can expect to live to age 80.

35. The death rate for a 50-year-old is about 5 per 1000, which means the probability that a randomly selected 50-year-old will die in the next year is 5/1000. Those that don't die will file no claim; the beneficiaries of those that do will file a $50,000 claim. Thus the expected value of a single policy (from the point of view of the insurance company) is $200 × 1 + (−$50,000) × 5/1000 + ($0) × 995/1000 = −$50. The company insures 1 million 50-year-olds, and expects to lose $50 per policy, so it can expect a loss of $50 million.

FURTHER APPLICATIONS

37. a. In Utah, about $49,870 \div 365 = 137$ people were born per day.

    b. In Maine, about $13,861 \div 365 = 38$ people were born per day.

    c. The birth rate in Utah was $\dfrac{49,870}{2.3 \times 10^6} \times 10^3 = 21.7$ births per 1000 people.

    d. The birth rate in Maine was $\dfrac{13,861}{1.3 \times 10^6} \times 10^3 = 10.7$ births per 1000 people.

39. a. There were $(13.5/1000) \times 300$ million $= 4.05$ million births in the U.S.

    b. There were $(8.1/1000) \times 300$ million $= 2.43$ million deaths in the U.S.

    c. The population increased by 4.05 million $-$ 2.43 million $= 1.62$ million people.

    d. About 3.0 million $-$ 1.62 million $= 1.38$ million people immigrated to the U.S. (this estimate neglects to account for emigration, though the number of people who emigrate from the U.S. is relatively low). The proportion of the overall population growth due to immigration was $1.38/3.0 = 46\%$.

41. a. The ratio of twins to people is about $125,000 \div 4,00,000 = 0.031$.

    b. The ratio of triplets (or higher) to people is about $7,000 \div 4,000,000 = 0.0018$.

UNIT 7E

QUICK QUIZ

1. **b**. This is an arrangement with repetition, with 36 choices at each selection.

3. **c**. The order of selection is important when counting the number of ways to arrange the roles.

5. **b**. The number of permutation of n items taken r at a time is always at least as large as the number of combinations of n items taken r at a time (by a factor of $r!$), and $_{15}P_7 = 32,432,400$ is much larger than $_7P_7 = 5040$.

7. **c**. The number of combinations of 9 items taken 4 at a time is $_9C_4 = \dfrac{9!}{(9-4)!4!} = \dfrac{9 \times 8 \times 7 \times 6}{4 \times 3 \times 2 \times 1}$.

9. **c**. The probability that it will be you is $1/100,000 = 0.00001$, so the probability it will not be you is $1 - 0.00001 = 0.99999$.

DOES IT MAKE SENSE?

15. Makes sense. The permutations formula is used when order of selection is important.

17. Makes sense. The number of such batting orders is $_{25}P_9 = 7.4 \times 10^{11}$, which is nearly 1 trillion ways. Clearly there is no hope of trying all of them.

19. Makes sense. This illustrates the general principle that *some* coincidence is far more likely that a *particular* coincidence.

BASIC SKILLS & CONCEPTS

21. $6! = 720$

23. $5! \div 3! = 20$.

25. $\dfrac{12!}{4!3!} = 3,326,400$.

27. $\dfrac{11!}{3!(11-3)!} = 165$.

29. $\dfrac{8!}{3!(8-3)!} = 56$.

31. $\dfrac{6!8!}{4!5!} = 10,800$.

33. This is an arrangement with repetition, where we have 10 choices (the digits 0–9) for each of 5 selections, and thus there are $10^5 = 100,000$ possible zip codes.

35. This is a permutation of 36 items taken 4 at a time because the order in which the characters are arranged is important (a different order results in a different password). Thus there are $_{36}P_4 = 1,413,720$ passwords.

37. This is a permutation of 8 items taken 8 at a time, and $_8P_8 = 40,320$.

39. Because order of selection does not matter, this is a combination of 25 items taken 6 at a time, or $_{25}C_6 = 177,100$.

41. Order matters, so this is a permutation of 6 items taken 6 at a time, which is $_6P_6 = 720$.

43. This is an arrangement with repetition, so use the multiplication principle to find that there are $26^3 \times 10^4 = 175,760,000$ such license plates.

45. This is an arrangement with repetition, so use the multiplication principle to get $9 \times 10 \times 10 \times 9 \times 10 \times 10 \times 10 \times 10 \times 10 \times 10 = 8,100,000,000$. Note that with the first digit and the fourth digit, there are only 9 choices.

47. If we assume that the letters can be repeated, this is an arrangement with repetition, and there are 4 choices for each of the 3 selections, which implies there are $4^3 = 64$ different "words."

49. The order in which the songs appear matters, so this is a permutation of 13 objects taken 13 at a time, and thus there are $_{13}P_{13} = 6,227,020,800$ different ways to order the songs.

51. Following Example 8 in the text, the probability that no one has your birthday is $1 - \left(\dfrac{364}{365}\right)^{11} = 0.0297$. The probability that at least one pair shares a birthday is given by

$$1 - \frac{364 \times 363 \times \ldots \times 354}{365^{11}} = 1 - \frac{1.276 \times 10^{28}}{1.532 \times 10^{28}} = 0.167.$$

FURTHER APPLICATIONS

53. a. Use the multiplication principle to get $20 \times 8 = 160$ different sundaes.

b. This is an arrangement with repetition, so there are $20^3 = 8000$ possible triple cones.

c. Since you specify the locations of the flavors, the order matters, and you should use permutations. There are $_{20}P_3 = 6840$ possibilities.

d. The order of the flavors doesn't matter, so use combinations. There are $_{20}C_3 = 1140$ possibilities.

55. This problem is best solved by trial and error. For Luigi, we know that 56 is the number of combinations of n items taken 3 at a time, and thus $_nC_3 = 56$. After a little experimentation, you will discover that $n = 8$; that is, Luigi uses 8 different toppings to create 56 different 3-topping pizzas. In a similar fashion, you can discover that Ramona uses 9 different toppings (solve $_nC_2 = 36$ for n).

57. There are $_{32}C_6 = 906,102$ ways to choose the balls, and only one way to match all six numbers, so the probability is 1/906,102.

59. There are $_{52}C_5 = 2,598,960$ five-card hands. Four of those consist of the 10, J, Q, K, and A of the same suit, so the probability is $4/2,598,960 = 1/649,740$.

61. There are $_{16}C_3 = 560$ ways to choose the top three spellers, and only one way to guess the correct three, so the probability is 1/560.

63. There are $_{52}C_5 = 2,598,960$ five-card hands. In order to get four-of-a-kind in, say, aces, you need to select all four aces, and then any of the other remaining 48 cards. Thus there are 48 ways to get a four-of-a-kind in aces. This is true for the other 12 card values as well, so there are $13 \times 48 = 624$ different four-of-a-kind hands. The desired probability is then $624/2,598,960 = 0.00024$.

65. a. The probability of winning five games in a row is $0.48^5 = 0.025$, assuming you play only five games. It is considerably higher if you play numerous games.

b. The probability of winning ten games in a row is $0.48^{10} = 0.00065$, assuming you play only ten games. It is considerably higher if you play numerous games.

c. Once again, assuming all 2000 people play only five games, one would expect about 2.5% of them would win five games in a row (because the probability of five-in-a-row is 0.025, as shown in part **a**). This is $0.025 \times 2000 = 50$ people.

d. Assuming all 2000 people play only ten games, one would expect about $0.00065 \times 2000 = 1.3$ people to win all ten games.

UNIT 8A

QUICK QUIZ

1. **b**. The absolute change in the population was 10,000 people, and if population grows linearly, the absolute change remains constant. Thus the population at the end of the second year will be 110,000 + 10,000 = 120,000.

3. **a**. Because your money is growing exponentially, if it doubles in the first 6 months, it will double every 6 months.

5. **c**. Exponential decay occurs whenever a quantity decreases by a constant percentage (over the same time interval).

7. **b**. Note that at 11:01, there are $2 = 2^1$ bacteria; at 11:02 there are $4 = 2^2$ bacteria; at 11:03 there are $8 = 2^3$, and so on. The exponent on the base of 2 matches the minute, and this pattern continues through 11:30, which means there are 2^{30} bacteria at that time.

9. **a**. As with the population, the volume of the colony doubles every minute, so it only takes one minute for the volume to double from 1 to 2 m^3.

DOES IT MAKE SENSE?

15. Makes sense. Any quantity that undergoes constant, positive percent change grows exponentially.

17. Makes sense. Any quantity that grows exponentially will eventually become very large, so as long as the small town's rate of growth is large enough, it can become a large city within a few decades.

BASIC SKILLS & CONCEPTS

19. The population is growing linearly because its absolute change remains constant at 623 people per year. The population in four years will be 2500 + 4 × 623 = 4992.

21. This is exponential growth because the cost of food in increasing at a constant 30% per year. Each year, the cost increases by a factor of 1.30, so your food bill was $120(1.30)^3 = \$263.64$.

23. The price is decreasing exponentially because the percent change remains at a constant −12% per year. Each year, the price decreases by a factor of 0.88, so its price in three years will be $\$50(0.88)^3 = \$34.07$.

25. The house's value is increasing linearly because the price increases by the same amount each year. In five years, it will be worth $100,000 + 5 × $15,000 = $175,000.

27. There should be $2^{14} = 16,384$ grains placed on the 15th square. The total number of grains would be $2^{15} - 1 = 32,767$, and it would weigh 32,767 × (1/7000) = 4.7 pounds.

29. When the chessboard is full, there are $2^{64} - 1$ grains, with total weight of $(2^{64} - 1)$ × (1/7000) = 2.64×10^{15} pounds, or about 1.3×10^{12} tons.

31. After 20 days, you would have $2^{20} = 1,048,576$ pennies, or $10,485.76.

33. After experimenting with various values of n in 2^n, you'll find that the balance has grown to about $1.4 billion after 37 days.

35. At 11:50, there are $2^{50} = 1.1 \times 10^{15}$ bacteria in the bottle. The bottle is full at 12:00 noon, so it was half-full at 11:59, 1/4-full at 11:58, 1/8-full at 11:57, and so on. Continuing in this manner, we find the bottle was 1/1024-full at 11:50. This could also be expressed as $1/2^{10}$-full at 11:50.

37. As discussed in the text, the volume of the colony of 2^{120} bacteria would occupy a volume of 1.3×10^{15} m^3. Divide this by the surface area of the earth to get an approximate depth:
$$1.3 \times 10^{15} \text{ } m^3 \div 5.1 \times 10^{14} \text{ } m^2 = 2.5 \text{ m}.$$
This is quite a bit more than knee-deep (it's more than 8 feet).

39. a. (Only 100-year intervals are shown here).

Year	Population
2000	6.0×10^9
2100	2.4×10^{10}
2200	9.6×10^{10}
2300	3.8×10^{11}
2400	1.5×10^{12}
2500	6.1×10^{12}
2600	2.5×10^{13}
2700	9.8×10^{13}
2800	3.9×10^{14}
2900	1.6×10^{15}
3000	6.3×10^{15}

b. In the year 2800, there would be 3.9×10^{14} people, each occupying one square meter, and in 2850, there would be 7.9×10^{14} people (again, each occupying one square meter). Since 5.1×10^{14} m^2 is in between these two values, it would happen between 2800 and 2850.

c. We would reach the limit when the earth's population reached $5.1 \times 10^{14} \div 10^4 = 5.1 \times 10^{10}$ people. This would happen shortly after the year 2150.

d. If by 2150 we've already reached the limit of 10^4 m^2 per person on earth, it will take only three more doubling periods to increase our need for space eight-fold. Since the doubling period is 50 years, this would take only 150 years, at which point we would have already exceeded the additional space available in the solar system (there is only five times more area out there).

UNIT 8B

QUICK QUIZ

1. **a.** The factor by which an exponentially growing quantity grows is $2^{t/T_{double}}$.

3. **a.** The approximate doubling time formula does not do a good job when the growth rate r goes higher than 15%.

5. **c.** During the first 12 years, half of the tritium decays, and during the second 12 years, half of that, or 1/4 decays.

7. **b.** The half-life can be approximated by $T_{half} = 70/r$.

9. **c.** The decimal equivalent of 20% is 0.2, though you should set $r = -0.2$ because the population is decreasing.

DOES IT MAKE SENSE?

19. Does not make sense. In 50 years, the population will increase by a factor of 4.

21. Does not make sense. While it's true that half will be gone in ten years, over the next ten years, only half of what remains will decay, which means there will still be 1/4 of the original amount.

BASIC SKILLS & CONCEPTS

23. True. $10^0 = 1$, and $10^1 = 10$, so $10^{0.928}$ should be between 1 and 10 (because 0.928 is between 0 and 1).

25. False. $10^{-5.2} = \frac{1}{10^{5.2}}$, which is quite small.

27. False. While it is true that π is between 3 and 4, $\log_{10} \pi$ is not (it's between 0 and 1).

29. False. $\log_{10}(10^6) = 6$, and $\log_{10}(10^7) = 7$. Because 1,600,000 is between 10^6 and 10^7, $\log_{10} 1,600,000$ is between 6 and 7.

31. True. Since $\log_{10}\left(\frac{1}{4}\right) = \log_{10}\left(4^{-1}\right) = -\log_{10} 4$, we know $\log_{10}\left(\frac{1}{4}\right)$ is between -1 and 0 because $\log_{10} 4$ is between 0 and 1.

33. a. $\log_{10} 8 = \log_{10}(2^3) = 3\log_{10} 2 = 3 \times 0.301 = 0.903$.

b. $\log_{10} 2000 = \log_{10} 2 + \log_{10} 10^3 = 0.301 + 3 = 3.301$.

c. $\log_{10} 0.5 = \log_{10} 1 - \log_{10} 2 = 0 - 0.301 = -0.301$.

d. $\log_{10} 64 = \log_{10} 2^6 = 6\log_{10} 2 = 6 \times 0.301 = 1.806$.

e. $\log_{10}\left(\frac{1}{8}\right) = \log_{10} 2^{-3} = -3\log_{10} 2 = -3 \times 0.301 = -0.903$.

f. $\log_{10} 0.2 = \log_{10}\left(\frac{2}{10}\right) = \log_{10} 2 - \log_{10} 10 = 0.301 - 1 = -0.699$.

35. In 12 hours, the fly population will increase by a factor of $2^{12/4} = 2^3 = 8$. Since one week is 168 hours, the population will increase by a factor of $2^{168/4} = 2^{42} = 4.4 \times 10^{12}$ in one week.

37. The population will increase by a factor of four (quadruple) in 30 years because $2^{30/15} = 4$.

39. In 12 years, the population will be $3500 \times 2^{12/10} = 8041$. In 24 years, it will be $3500 \times 2^{24/10} = 18,473$.

41. Note that 3 years is 36 months, and 4 years is 48 months. In 3 years, the number of cells will be $1 \times 2^{36/2.5} = 21,619$. In 4 years, the number will be $1 \times 2^{48/2.5} = 602,249$.

43. The year 2016 is 10 years later than 2006, so the population will be $6.5 \times 2^{10/45} = 7.6$ billion. In 2056 (50 years later), the population will be $6.5 \times 2^{50/45} = 14.0$ billion. In 2106 (100 years later), it will be $6.5 \times 2^{100/45} = 30.3$ billion.

45.

Month	Population
0	100
1	$100 \times (1.07)^1 = 107$
2	$100 \times (1.07)^2 = 114$
3	$100 \times (1.07)^3 = 123$
4	$100 \times (1.07)^4 = 131$
5	$100 \times (1.07)^5 = 140$
6	$100 \times (1.07)^6 = 150$
7	$100 \times (1.07)^7 = 161$
8	$100 \times (1.07)^8 = 172$
9	$100 \times (1.07)^9 = 184$
10	$100 \times (1.07)^{10} = 197$
11	$100 \times (1.07)^{11} = 210$
12	$100 \times (1.07)^{12} = 225$
13	$100 \times (1.07)^{13} = 241$
14	$100 \times (1.07)^{14} = 258$
15	$100 \times (1.07)^{15} = 276$

Because the population after ten months is 197, which is almost twice the initial population of 100, the doubling time is just over 10 months. The approximate doubling time formula claims that $T_{\text{double}} = 70/7 = 10$ months, which is pretty close to what we can discern from the table. Neither answer is exact, though.

47. $T_{\text{double}} \sim 70/4 = 17.5$ years. Prices will increase by a factor of $(1.04)^3 = 1.12$ in three years. Using the doubling time formula, prices would increase by a factor of $2^{3/17.5} = 1.13$, so the approximate doubling time formula does a reasonable job.

49. $T_{\text{double}} \sim 70/0.9 = 78$ months. Prices will increase by a factor of $(1.009)^{12} = 1.11$ in one year, and by a factor of $(1.009)^{96} = 2.36$ in eight years. Using the doubling time formula, prices would increase by a factor of $2^{12/78} = 1.11$ in one year, and by a factor of $2^{96/78} = 2.35$ in eight years, so the approximate doubling time formula does a reasonable job.

51. After 140 years, the fraction of radioactive substance that remains is $\left(\dfrac{1}{2}\right)^{140/70} = \dfrac{1}{4}$. After 280 years, the fraction remaining is $\left(\dfrac{1}{2}\right)^{280/70} = \dfrac{1}{16}$.

53. The concentration of the drug decreases by a factor of $\left(\dfrac{1}{2}\right)^{24/18} = 0.40$ after 24 hours, and by a factor of $\left(\dfrac{1}{2}\right)^{36/18} = \dfrac{1}{4}$ after 36 hours.

55. In 40 years, the population will be $1,000,000 \times \left(\dfrac{1}{2}\right)^{40/20} = 250,000$, and it will be $1,000,000 \times \left(\dfrac{1}{2}\right)^{70/20} = 88,388$ in 70 years.

57. In 150 days, $1 \times \left(\dfrac{1}{2}\right)^{150/77} = 0.26$ kg of cobalt will remain, and $1 \times \left(\dfrac{1}{2}\right)^{300/77} = 0.067$ kg of cobalt will remain after 300 days.

59. $T_{\text{half}} \sim 70/8 = 8.75$ years. The fraction of the forest that will remain after 30 years is $(1 - 0.08)^{30} = 0.082$. Using the half-life formula, the fraction that will remain is $\left(\dfrac{1}{2}\right)^{30/8.75} = 0.093$, so the approximate half-life formula is doing a reasonable job.

61. $T_{\text{half}} \sim 70/9 = 7.8$ years. After 50 years, the population will be $10,000(1 - 0.09)^{50} = 90$ elephants. Using the half-life formula, the population will be $10,000 \times \left(\dfrac{1}{2}\right)^{50/7.8} = 118$ elephants, so the approximate half-life formula is doing a reasonable job.

63. The approximate doubling time is $T_{\text{double}} \sim 70/12 = 5.8$ years. (A rounded value for) the exact doubling time is $\dfrac{\log_{10} 2}{\log_{10}(1 + 0.12)} = 6.11$ years. The price of a \$500 item in four years will be $\$500 \times 2^{4/6.11} = \$787.12$. (Note that this answer is using the rounded value for the "exact" doubling time – if you use all the digits your calculator stores, you should get a price of \$786.76).

65. The approximate doubling time is $T_{\text{double}} \sim 70/4 = 17.5$ years. (A rounded value for) the exact doubling time is $\dfrac{\log_{10} 2}{\log_{10}(1 + 0.04)} = 17.67$ years. The nation's population in 30 years will be $1,000,000 \times 2^{30/17.67} = 324,404,283$. (Note that this answer is using the rounded value for the "exact" doubling time – if you use all the digits your calculator stores, you should get a population of 324,339,751).

FURTHER APPLICATIONS

67. The amount of plutonium found today would be $10^{12} \times \left(\dfrac{1}{2}\right)^{4.6 \times 10^9 / 24,000} = 0$ tons. The reason there is no plutonium left is that 4.6 billion ÷ 24,000 is nearly 200,000, which means whatever amount of plutonium was originally present has undergone about 200,000 halving periods, and that's enough to reduce any amount imaginable to 0 (even if the entire mass of the earth were radioactive plutonium at the outset, it would have all decayed to other elements in 4.6 billion years).

UNIT 8C

QUICK QUIZ

1. **c.** Convert 75 million people per year into people per minute: $\dfrac{75 \times 10^6 \text{ people}}{\text{yr}} \times \dfrac{1 \text{ yr}}{365 \text{ d}} \times \dfrac{1 \text{ d}}{24 \text{ h}} \times \dfrac{1 \text{ hr}}{60 \text{ min}}$ = 143 people per minute (or about 140 people/min).

3. **b.** The birth rate peaked in 1960, and has seen a significant decline since then, but it is not the primary reason for the rapid growth – the decreasing death rate has played the most important role over the last three centuries.

5. c. Answers **a** and **b** would tend to decrease the carrying capacity, while developing a cheap source of energy would allow us to solve some of the barriers to population growth, such as clean water for all.

7. c. The logistic model looks like exponential growth at the outset, but then the growth rate declines, and the population levels out to a level at (or below) the carrying capacity.

9. a. The approximate doubling time of a population growing at 1.2% per year is 70/1.2 = 58.3 years.

DOES IT MAKE SENSE?

17. Makes sense. If the current growth rate of 1.2% remains steady for the next decade, we can expect a population of $6.5 \times 10^9 (1.012)^{10} = 7.3$ billion, which is an increase of 800 million people (i.e. more than twice the size of the current U.S. population of 300 million).

19. Does not make sense. The carrying capacity depends upon many factors, such as our ability to produce food for the current population.

21. Does not make sense. Predator/prey models often include the idea of overshoot and collapse.

BASIC SKILLS & CONCEPTS

23. $T_{\text{double}} \sim 70/0.9 = 78$ years. Under this assumption, the population in 2050 would be $6 \times 2^{50/78} = 9.4$ billion.

25. $T_{\text{double}} \sim 70/1.6 = 44$ years. Under this assumption, the population in 2050 would be $6 \times 2^{50/44} = 13.2$ billion.

27. a. The growth rate in 1985 was 14.5 – 11.8 = 2.7 per 1000 people. In 1995, it was 9.3 – 11.4 = –2.1 per 1000 people. In 2003, it was 9.0 – 11 = –2.0 per 1000 people.
   b. The birth rate decreased while the death rate remained fairly constant, which results in a declining growth rate. In this case, the growth rate became negative, which implies a declining population. If these trends continue, we can expect a decreasing population in the Czech Republic, though because there are so many factors that influence the growth rate of a country, it would be unwise to put much faith in these predictions 20 years hence.

29. a. The growth rate in 1985 was 11.8 – 11.3 = 0.5 per 1000 people. In 1995, it was 11.7 – 11.0 = 0.7 per 1000 people. In 2003, it was 11.0 – 10.0 = – 1.0 per 1000 people.
   b. The birth rate decreased, as did the death rate, though the difference between the two was small (but positive), which results in a slowly growing population. If these trends continue, we can expect a slowly growing population in Sweden, though

because there are so many factors that influence the growth rate of a country, it would be unwise to put much faith in these predictions 20 years hence.

31. When the population is 10 million, the actual growth rate is $4\% \times \left(1 - \dfrac{10,000,000}{60,000,000}\right) = 3.3\%$. When the population is 30 million, the actual growth rate is $4\% \times \left(1 - \dfrac{30,000,000}{60,000,000}\right) = 2.0\%$. When the population is 50 million, the actual growth rate is $4\% \times \left(1 - \dfrac{50,000,000}{60,000,000}\right) = 0.67\%$.

FURTHER APPLICATIONS

33. $T_{\text{double}} \sim 70/0.7 = 100$ years. In 2056, the population will be $3 \times 10^8 \times 2^{50/100} = 424$ million. In 2106, the population will be $3 \times 10^8 \times 2^{100/100} = 600$ million.

35. $T_{\text{double}} \sim 70/1.0 = 70$ years. In 2056, the population will be $3 \times 10^8 \times 2^{50/70} = 492$ million. In 2106, the population will be $3 \times 10^8 \times 2^{100/70} = 808$ million.

37. Answers will vary depending on your age.

39. We must first compute the base growth rate r in the logistic model, using the 1960s data:
$$r = \frac{2.1\%}{\left(1 - \dfrac{3 \text{ billion}}{8 \text{ billion}}\right)} = 3.36\%.$$
The current growth rate can now be predicted using a population of 6.5 billion:
$$\text{growth rate} = 3.36\% \times \left(1 - \frac{6.5 \text{ billion}}{8 \text{ billion}}\right) = 0.63\%.$$
This is lower than the actual current growth rate of 1.1%, which indicates the assumed carrying capacity of 8 billion is too low to fit well with the data from the 1960s to present.

41. We must first compute the base growth rate r in the logistic model, using the 1960s data:
$$r = \frac{2.1\%}{\left(1 - \dfrac{3 \text{ billion}}{15 \text{ billion}}\right)} = 2.625\%.$$
The current growth rate can now be predicted using a population of 6.5 billion:
$$\text{growth rate} = 2.625\% \times \left(1 - \frac{6.5 \text{ billion}}{15 \text{ billion}}\right) = 1.5\%.$$
This is higher than the actual current growth rate of 1.1%, which indicates the assumed carrying capacity of 15 billion is too high to fit well with the data from the 1960s to present.

43. With an annual growth rate of 2%, the approximate doubling time would be $T_{\text{double}} \sim 70/2 = 35$ years. The population of the city in the year 2000 (ten years beyond 1990) would have been

about $100,000(2)^{10/35} = 122,000$. The predicted population in 2010 and 2050 would be 149,000 and 328,000, respectively (the computation is similar to the 2000 prediction). With an annual growth rate of 5%, the approximate doubling time would be $T_{double} \sim 70/5 = 14$ years. The population of the city in the year 2000 would have been $100,000(2)^{10/14} = 164,000$. The predicted population in 2010 and 2050 would be 269,000 and 1,950,000, respectively.

UNIT 8D

QUICK QUIZ

1. **a**. Each magnitude on the Richter scale represents about 32 times as much energy as the previous magnitude.

3. **a**. By definition, a sound of 0 dB is the softest sound audible by a human.

5. **a**. From the definition of the decibel scale, we have $\dfrac{\text{intensity of sound}}{\text{intensity of softest audible sound}} = 10^{(10/10)} = 10$.

   The ratio on the left tells us a 10-dB sound is ten times as large as the softest audible sound, which is 0 dB.

7. **c**. On the pH scale, a lower number corresponds to higher acidity.

9. **c**. Because $pH = -\log_{10}[H^+]$, we have $pH = -\log_{10} 10^{-5} = -(-5) = 5$.

DOES IT MAKE SENSE?

15. Does not make sense. Each unit on the Richter scale represents an increase of nearly 32 fold in the amount of energy released, and thus an earthquake of magnitude 8 releases about 32^4 (close to 1 million) times as much energy as an earthquake of magnitude 4. This does not imply that a magnitude 8 earthquake will do 1 million times as much damage, but it's not likely it will do only twice the damage of a magnitude 4 quake.

17. Does not make sense. The pH of water is related to its hydrogen ion concentration, which is not affected by its volume.

BASIC SKILLS & CONCEPTS

19. $E = (2.5 \times 10^4) \cdot 10^{1.5 \times 6} = 2.5 \times 10^{13}$ joules.

21. $E = (2.5 \times 10^4) \cdot 10^{1.5 \times 9} = 7.9 \times 10^{17}$ joules.

23. The energy released by a magnitude 6 quake is 2.5×10^{13} joules (see exercise 19), and since $5 \times 10^{15} \div 2.5 \times 10^{13} = 200$, the bomb releases 200 times as much energy.

25. As shown in Table 8.5, ordinary conversation is around 60 dB, and is $10^6 = 1,000,000$ times as loud as the softest audible sound.

27. The loudness is $10 \log_{10}(20 \times 10^6) = 73$ dB.

29. As with Example 4 in the text, we can compare the intensities of two sounds using the following: $\dfrac{\text{intensity of sound 1}}{\text{intensity of sound 2}} = 10^{(45-10)/10} = 3162$.

    Thus a 45-dB sound is 3162 times as loud as a 10-dB sound.

31. The distance has decreased by a factor of 5, which means the intensity of sound will increase by a factor of $5^2 = 25$.

33. The distance has decreased by a factor of 2, which means the intensity of sound will increase by a factor of $2^2 = 4$.

35. Each unit of increase on the pH scale corresponds to a decrease by a factor of 10 in the hydrogen ion concentration. Thus an increase of 3 on the pH scale corresponds to a decrease in the hydrogen ion concentration by a factor of $10^3 = 1000$. This makes the solution more basic.

37. $[H^+] = 10^{-7.5} = 3.2 \times 10^{-8}$ mole per liter.

39. $pH = -\log_{10}(0.01) = 2$. This is an acid, as is any solution with pH less than 7.

41. Acid rain (pH = 2) is $10^4 = 10,000$ times more acidic than ordinary water (pH = 6). See exercise 35.

FURTHER APPLICATIONS

43. According to Table 8.4, earthquakes with magnitudes between 2 and 3 are labeled "very minor," and they occur about 1000 times per year worldwide. The glasses in the cupboards of the Los Angeles residents near the epicenter may rattle around a bit, but nothing serious would be expected to happen.

45. A solution with a pH of 12 is very basic: 100,000 times more basic than pure water. When children ingest such liquids, the mouth, throat, and digestive tracts are often burned. One would hope that a secondary effect of the incident would be that an adult would call the poison center immediately.

47. You probably would not be able to hear your friend (the noise of the sirens is louder). If your friend is shouting: "Watch out for that falling piano above you!" you might die.

49. a. The distance has increased by a factor of 100, so the intensity of the sound will decrease by a factor of $100^2 = 10,000$ times (this is due to the inverse

square law). According to Table 8.5, busy street traffic is 80 dB, and is 10^8 times as loud as the softest audible sound. Note that a 40-dB sound is 10^4 times as loud as the softest sound, and thus is 10,000 times less intense than an 80-dB sound.

b. As with Example 4 in the text, we can compare the intensities of two sounds using the following:

$$\frac{\text{intensity of sound 1}}{\text{intensity of sound 2}} = 10^{(135-120)/10} = 31.6 .$$

Thus a 135-dB sound is 31.6 times as loud as a 120-dB sound. In order to reduce the intensity of the sound by a factor of 31.6, you need to increase the distance from the speaker by a factor of $\sqrt{31.6}$ = 5.6, so you should sit 5.6×10 m = 56 meters away.

c. The distance from you to the booth (8 meters) is 8 times as much as the distance between the people talking (1 meter), so the intensity of the sound will have decreased by a factor of 64 by the time it reaches you. A 20-dB sound is 100 times louder than the softest audible sound, and when decreased by a factor of 64, it becomes (100/64) times as loud as the softest sound. Therefore its loudness in decibels is $10 \log_{10}\left(\frac{100}{64}\right)$ = 1.94 dB.

Note that a 60-dB sound is 10,000 times as loud as a 20-dB sound (Table 8.5). If you wanted to amplify the sound that reaches you to 60 dB, you would need to increase it by a factor of 64 to overcome the loss of intensity due to distance (this would bring it back to 20 dB), and increase it again by a factor of 10,000 to raise it to the 60-dB level. This is equivalent to amplifying the sound by a factor of 640,000. Pray that the waiter doesn't come to your table and say a loud "hello" while your earpiece is in.

51. a. $[H^+] = 10^{-4}$ mole per liter.

b. The lake water with pH = 7 has a hydrogen ion concentration of 10^{-7} mole per liter. There are 3.785 liters in a gallon, so a 100 million gallon lake has 378,500,000 = 3.785×10^8 liters in it. Multiplying the hydrogen ion concentration by the volume of the lake gives us 37.85 moles of hydrogen ions before the acid was added. In a similar fashion, we find that 100,000 gallons (378,500 liters) of acid with pH = 2 ($[H^+] = 10^{-2}$) has 3785 moles of hydrogen ions. Combining both the volumes of the liquids and the moles of hydrogen ions gives us a hydrogen ion concentration of $\frac{37.85 + 3785 \text{ mole}}{3.785 \times 10^8 + 378,500 \text{ L}} = 10^{-5}$ mole per liter. This implies the polluted lake has pH = $-\log_{10}(10^{-5})$ = 5.

c. Proceeding as in part **b**, the acid-rain lake (without added acid) has 37,850 moles of hydrogen ions. When the acid is added (increasing the number of moles by 3785), the hydrogen ion concentration becomes $\frac{37,850 + 3785 \text{ mole}}{3.785 \times 10^8 + 378,500 \text{ L}} =$ 1.1×10^{-4} mole per liter. This implies the polluted lake has pH = $-\log_{10}(1.1 \times 10^{-4})$ = 3.96.

d. The pollution could be detected if the acid was dumped into a lake with pH = 7 (part **b**), but it could not be detected if it was dumped into a lake with pH = 4 (part **c**).

UNIT 9A

QUICK QUIZ

1. **a.** A function describes how a dependent variable changes with respect to one or more independent variables.

3. **c.** The DJIA changes with respect to the passage of time, and thus is a function of time.

5. **b.** The dependent variable is normally plotted along the vertical axis, and in this case, the dependent variable is z (with an independent variable of w).

7. **a.** The speed of the car is the independent variable (because gas mileage changes with respect to speed), and the domain of a function is the set of values of interest for the independent variable.

9. **c.** Traffic volume on an urban freeway peaks in the morning rush hour, peaks again in the evening rush hour, and typically repeats this pattern day after day. If weekend traffic were significantly different from the weekday pattern, we would still see a weekly pattern repeating itself – this is a hallmark of a periodic function.

DOES IT MAKE SENSE?

15. Makes sense. Climatologists use mathematical models to predict and understand the nature of the earth's climate.

17. Does not make sense. If your heart rate depends upon your running speed, heart rate is the dependent variable, and the range is used to describe the values of the dependent variable.

BASIC SKILLS & CONCEPTS

19.

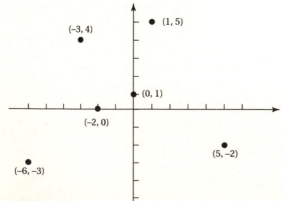

21. As the volume of the tank increases, the cost of filling it also increases.

23. The speed of the skydiver increases with time until reaching constant terminal velocity. This describes the relationship of a skydiver falling a sufficient distance before opening his chute. When the chute is opened, the speed decreases to a different terminal velocity until the skydiver reaches the ground.

25. As the average speed of the car increases, the travel time decreases.

27. As the gas mileage increases, the cost of driving a fixed distance decreases (assuming that gas costs remain constant).

29. a. The pressure at 6000 feet is about 24 inches of mercury; the pressure at 18,000 feet is about 17 inches; and the pressure at 29,000 feet is about 11 inches.

b. The altitude is about 8000 feet when the pressure is 23 inches of mercury; the altitude is about 14,000 feet when the pressure is 19 inches; and the altitude is about 25,000 feet when the pressure is 13 inches.

c. It appears that the pressure reaches 5 inches of mercury at an altitude of about 50,000. The pressure approaches zero as one moves out of earth's atmosphere and into outer space.

31. a. The independent variable is time (in years), and the dependent variable is world population. The domain is the set of years from 1950 to 2000, and the range is the set of population values between roughly 2.5 billion and 6 billion people.

b. The function shows a steadily increasing world population between 1950 and 2000.

33. a. The variables are (*time, temperature*), or (*date, temperature*). The domain is all days over the course of a year, and the range is the set of temperatures between 38°F and 85°F.

b.

c. The temperature increases during the first half of the year, and then decreases during the second half of the year.

35. a. The variables are (*year, tobacco production*). The domain is the set of years between 1975 and 1990, and the range is the set of weights between 1.2 and 2.2 billion pounds.

b.

c. In this case, because the data consist of yearly figures for production, it would not be appropriate to fill in the graph between the data points. The function, then, is simply the production numbers associated with each year.

37. a. The domain of the function (*altitude, temperature*) is the set of altitudes of interest – say, 0 ft to 15,000 ft (or 0 m to 4000 m). The range is the set of temperatures associated with the altitudes in the domain; the interval 30°F to 90°F (or about 0°C to 30°C) would cover all temperatures of interest (this assumes a summer ascent of a mountain in the temperate zone of the earth).
b.

39. a. The domain of the function (*blood alcohol content, reflex time*) consists of all reasonable BACs (in gm/100 mL). For example, numbers between 0 and 0.25 would be appropriate. The range would consist of the reflex times associated with those BACs.
b.

41. a. The domain of the function (*time of day, traffic flow*) consists of all times over a full day. The range consists of all traffic flows (in units of cars per minute) at the various times of the day. We would expect light traffic flow at night, medium traffic flow during the midday hours, and heavy traffic flow during the two rush hours.

41. b.

43. a. The domain of the function (*time, world record in the 100-meter dash*) consists of all years from 1970 to 2004. The range consists of the world records for the race at all times between 1970 and 2004. Some research would be needed to find the exact records and the days on which they were lowered. The graph shows a slow decrease (from about 10.2 seconds to 9.8 seconds for men, and from about 11.0 seconds to 10.4 seconds for women) over this period of time.
b.

45. a. The domain of the function (*time, population of China*) is all years from 1900 to, say, 2010. The range consists of the population values of China during those years (roughly 400 million to 1.3 billion).
b.

47. a. The domain of the function (*angle of cannon, horizontal distance traveled by cannonball*) is all cannon angles between 0° and 90°. The range would consist of all horizontal distances traveled by the cannonball for the various angles in the domain. It is well known that a projectile has maximum range when the angle is about 45°, so the graph shows a peak at 45°.
b.

UNIT 9B

QUICK QUIZ

1. **c.** Constant absolute change over the same time interval (or other variable) gives rise to linear models, and this amounts to constant slope.

3. **c.** When a line slopes downward, the ratio of its change in y to change in x is negative, and this amounts to a negative slope or negative rate of change.

5. **c.** Larger rates of change correspond with larger and steeper slopes.

7. **c.** The coefficient on the independent variable is the slope, which also represents the rate of change. Since this coefficient is negative, we can expect a graph that slopes downward.

9. **b.** A line in the form $y = mx + b$ has slope $= m$ and y-intercept $= b$, so the slope of both $y = 12x - 3$ and $y = 12x + 3$ is $m = 12$, and they have different y-intercepts (-3 and 3).

DOES IT MAKE SENSE?

17. Does not make sense. The graph of a linear function is always a line.

19. Makes sense. The familiar relationship *distance* = *speed* × *time* can be solved for *speed* to produce *speed* = *distance* ÷ *time*, which is the rate of change in distance with respect to time.

BASIC SKILLS & CONCEPTS

21. a. Rain depth increases linearly with time.
    b. It appears that the change in depth over the first three hours is four inches, so the slope is 4/3 inches per hour.
    c. The model is realistic if the rainfall rate is a constant 4/3 inches per hour over four hours.

23. a. On a long trip, the distance from home decreases linearly with time.
    b. The distance has decreased by 500 miles over the span of 7 hours, so the slope is $-500 \div 7 = -71.4$ miles per hour.
    c. This is a good model if the speed of travel is a constant 71.4 miles per hour over 7 hours.

25. a. Shoe size increases linearly with the height of the individual.
    b. The change in shoe size was 11 while the change in height was 80 inches, and this produces a slope of $11 \div 80 = 0.1375$ size per inch.
    c. This model is a rough approximation at best, and it would be difficult to find realistic conditions where the relationship held.

27. The *water depth* decreases with respect to *time* at a rate of 1.5 in./day, so the rate of change is -1.5 in./day. In 6.5 days, the water depth decreases by

1.5 in./day × 6.5 days = 9.75 inches. In 12.5 days, it decreases by 1.5 in./day × 12.5 days = 18.75 in.

29. The *height* of the plant increases with respect to *time* at a rate of 0.8 in./day. In 30 days, the height increases by 0.8 in./day × 30 days = 24 inches. In 60 days, the height increases 0.8 in./day × 60 days = 48 inches.

31. The *snow depth* increases with respect to *time* at a rate of 3.5 inches per hour. In 6.3 hours, the snow depth will have increased by 3.5 in./hr × 6.3 hr = 22.05 inches. In 9.8 hours, the snow depth will have increased by 3.5 in./hr × 9.8 hr = 34.3 in.

33. The independent variable is time t, measured in years, where $t = 0$ represents today. The dependent variable is price p. The equation for the price function is $p = 15,000 + 800t$. A new car will cost $p = 15,000 + 800 \times 3.5 = \$17,800$. This function does not give a good model for car prices.

35. The independent variable is snow depth d, measured in inches, and the dependent variable is maximum speed s, measured in miles per hour. The equation for the speed function is $s = 40 - 1.1d$. To find the depth at which the plow will not be able to move, set $s = 0$ and solve for d.
$$0 = 40 - 1.1d$$
$$1.1d = 40$$
$$d = 40/1.1 = 36 \text{ inches.}$$
The rate at which the speed decreases per inch of snow depth is probably not a constant, so this model is an approximation.

37. The independent variable is time t, measured in minutes, and the dependent variable is rental cost r. As long as the copy business is willing to prorate rental charges per minute (rather than in 5-minute blocks), the change per minute is $1.50/5 = \$0.30$ per minute. This implies the rental cost function is $r = 8 + 0.30t$. To find out how many minutes can be rented for $25, set $r = 25$, and solve for t.
$$25 = 8 + 0.30t$$
$$17 = 0.30t$$
$$t = 17/0.30 = 56.7 \text{ minutes.}$$
This function gives a very good model of rental costs.

39. Let W represent the weight of the dog in pounds, and t the time in years. The slope of the model is $(15 - 2.5) \div (1 - 0) = 12.5$ pounds per year, so the linear function is $W = 12.5t + 2.5$. When the dog is 5 years old, its weight is $W = 12.5 \times 5 + 2.5 = 65$ pounds, and at age 10, it weighs $W = 12.5 \times 10 + 2.5 = 127.5$ pounds. The model is accurate only for small ages.

41. Let P represent the profit (or loss) realized when selling n raffle tickets. The initial cost of the raffle to the fundraisers is $350 (a negative profit), and the rate of change of the profit with respect to the

number of tickets sold is $10 per ticket. Thus the profit function is $P = 10n - 350$. One can see by inspection that 35 tickets must be sold to break even.

43. Let V represent the value of the washing machine, and let t represent time (in years). The value function is $V = -75t + 1200$. Set $V = 0$ and solve for t to find out how long it will take for the value to reach $0.

$$0 = -75t + 1200$$
$$75t = 1200$$
$$t = 1200/75 = 16 \text{ years.}$$

FURTHER APPLICATIONS

45. The slope is $m = 2$, and the y-intercept is $(0, 6)$.

47. The slope is $m = -5$, and the y-intercept is $(0, -5)$.

49. The slope is $m = 3$, and the y-intercept is $(0, -6)$.
51. The slope is $m = -1$, and the y-intercept is $(0, 4)$.
Graphs for Exercises 49–52:

53. The independent and dependent variables are *time* and *elevation*, respectively.

After 3.5 hours, the elevation of the climbers is $6500 + 600 \times 3.5 = 8600$ feet. The model is reasonable provided the rate of ascent is nearly constant.

55. The independent and dependent variables are *number of posters* and *cost*, respectively.

The cost of producing 2000 posters is $2000 + 2000 \times 3 = \$8000$. This function probably gives a fairly realistic estimate of printing costs.

57. The independent and dependent variables are *time* and *cost*, respectively.

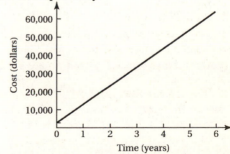

The cost of 6 years of school is $2000 + 10,000 \times 6 = \$62,000$. Provided costs do not change during the six-year period, this function is an accurate model of the cost.

59. a. Half of the fish caught in the second outing are tagged, and this means the proportion of those tagged in the entire population is 1/2. Thus $200/N = 1/2$, which implies $N = 400$ fish.

b. One-fourth of the fish caught in the second outing are tagged, and this means the proportion of those tagged in the entire population is 1/4. Thus $200/N = 1/4$, which implies $N = 800$ fish.

c. If p represents the proportion of those caught in the second outing that are tagged, we have $200/N = p$, which implies $N = f(p) = 200/p$.

d. Graph not supplied, though it looks similar to $y = 1/x$, with domain of $0 < p \leq 1$. When p is small, N is very large, and when p is close to 1, N is close to 200. Note that the above domain is the domain of the mathematical model, not the set of all possible values of p that could be observed in such a study. Consider a situation where only 200 fish are caught in the first outing from a very large lake with a large fish population. On the second outing, you may find that none of the 200 fish caught is tagged, in which case $p = 0$. This makes sense in the field, but it does not make sense in the mathematical model, because we cannot predict the fish population by using $N = 200/0$ (division by 0 is not allowed).

e. If $p = 15\% = 0.15$, $N = 200/0.15 = 1333$ fish.

UNIT 9C

QUICK QUIZ

1. **a.** Q_0 represents the initial value in the exponential model.

3. **c.** Because the value of the dollar is decreasing, r is negative.

5. **c.** If you start out with a positive quantity, and remove half of it, half still remains (and it is a positive quantity, which can never equal 0).

7. **c.** You need to know both the initial amount Q_0 and the amount at some future time t in order to solve the exponential model $Q = Q_0 \times \left(\frac{1}{2}\right)^{t/T_{half}}$ for t (you must also know T_{half}, but we are given that).

9. **a.** When you solve $\frac{1}{25}Q_0 = Q_0 \times \left(\frac{1}{2}\right)^{t/700,000,000}$ for t, you get the expression given in answer **a**.

DOES IT MAKE SENSE?

17. Does not make sense. After 100 years, the population growing at 2% per year will have increased by a factor of $1.02^{100} = 7.245$, whereas the population growing at 1% per year will have increased by a factor of $1.01^{100} = 2.705$. Under only very special circumstances will the first population grow by twice as many people as the second population. (Let Q_0 represent the initial amount of the population growing by 2%, and R_0 the initial amount of the population growing by 1%. After 100 years, Q_0 will have grown to $7.245Q_0$, and thus will have grown by $7.245Q_0 - Q_0 = 6.245Q_0$. In a similar fashion, it can be shown that R_0 will have grown by $1.705R_0$. We need $6.245Q_0 = 2 \times 1.705R_0$ in order to satisfy the condition that the first population will grow by twice as many people as the second population. This implies that the ratio $R_0/Q_0 = 6.245/3.41 = 1.83$. In other words, whenever the initial amount of the second population is 1.83 times as large as the initial amount of the first population, the first population will grow by twice as many people in 100 years. Try it with initial populations of 100 and 183.)

19. Makes sense. If we know the half-life of the radioactive material, we can create an exponential function that models the quantity remaining as time passes. In order to use this model to determine ages of bones, we only need to measure the amount of radioactive material present in the bone, and the amount it had originally (the latter is usually surmised by making assumptions about conditions present at the time of death).

BASIC CONCEPTS & SKILLS

21. $2^x = 128 \Rightarrow \log_{10} 2^x = \log_{10} 128 \Rightarrow x\log_{10} 2 = \log_{10} 128 \Rightarrow x = \frac{\log_{10} 128}{\log_{10} 2} = 7$.

23. $3^x = 99 \Rightarrow \log_{10} 3^x = \log_{10} 99 \Rightarrow x\log_{10} 3 = \log_{10} 99 \Rightarrow x = \frac{\log_{10} 99}{\log_{10} 3} = 4.18$.

25. $7^{3x} = 623 \Rightarrow \log_{10} 7^{3x} = \log_{10} 623 \Rightarrow 3x\log_{10} 7 = \log_{10} 623 \Rightarrow x = \frac{\log_{10} 623}{3\log_{10} 7} = 1.10$.

27. $9^x = 1748 \Rightarrow x = \frac{\log_{10} 1748}{\log_{10} 9} = 3.40$. (See exercise 23 for a similar process of solution.)

29. $\log_{10} x = 4 \Rightarrow 10^{\log_{10} x} = 10^4 \Rightarrow x = 10,000$.

31. $\log_{10} x = 3.5 \Rightarrow x = 10^{3.5} = 3162.28$. (See exercise 29 for a similar process of solution).

33. $3\log_{10} x = 4.2 \Rightarrow \log_{10} x = 1.4 \Rightarrow x = 10^{1.4} = 25.12$.

35. $\log_{10}(4 + x) = 1.1 \Rightarrow 4 + x = 10^{1.1} \Rightarrow x = 8.59$.

37. a. $Q = 60,000 \times (1.025)^t$, where Q is the population of the town, and t is time (measured in years).

b.

Year	Population
0	60,000
1	61,500
2	63,038
3	64,613
4	66,229
5	67,884
6	69,582
7	71,321
8	73,104
9	74,932
10	76,805

c.

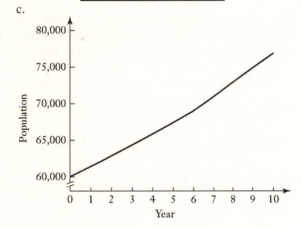

39. a. $Q = 1,000,000 \times (0.93)^t$, where Q is the number of acres of forest left after t years.

b.

Year	Acres ($\times 10^6$)
0	1.00
1	0.93
2	0.86
3	0.80
4	0.75
5	0.70
6	0.65
7	0.60
8	0.56
9	0.52
10	0.48

c.

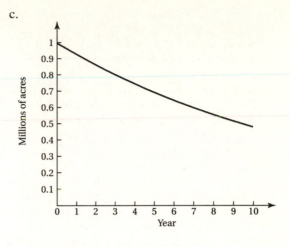

41. a. $Q = 175,000 \times (1.05)^t$, where Q is the average price of a home in dollars, and t is time, measured in years, with $t = 0$ corresponding to 2007.

b.

Year	Average price
0	$175,000
1	$183,750
2	$192,938
3	$202,584
4	$212,714
5	$223,349
6	$234,517
7	$246,243
8	$258,555
9	$271,482
10	$285,057

c.

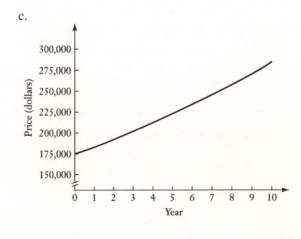

43. a. $Q = 2000 \times (1.05)^t$, where Q is your monthly salary in dollars, and t is time, measured in years.

b.

Year	Monthly Salary
0	$2000.00
1	$2100.00
2	$2205.00
3	$2315.25
4	$2431.01
5	$2552.56
6	$2680.19
7	$2814.20
8	$2954.91
9	$3102.66
10	$3257.79

c.

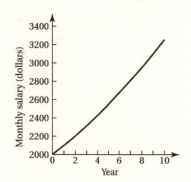

45. Over the span of a year, prices will increase by a factor of $(1.012)^{12} = 1.154$, which is a 15.4% increase.

47. Over the course of a year, prices would have risen by a factor of $(1.8)^{12} = 1156.83$, or an annual increase of 115,683%. Assuming one day is 1/30 of a month, prices would rise by a factor of $(1.8)^{(1/30)} = 1.0198$ in one day, which is a 1.98% increase.

49. An exponential model for the animal population is $Q = 1500 \times (0.92)^t$. We need to find the time t that will produce a population $Q = 30$ animals, which amounts to solving $30 = 1500 \times (0.92)^t$ for t.

$$\frac{30}{1500} = 0.92^t \Rightarrow 0.02 = 0.92^t \Rightarrow \log_{10} 0.02 =$$

$$\log_{10} 0.92^t \Rightarrow \log_{10} 0.02 = t \log_{10} 0.92 \Rightarrow$$

$$t = \frac{\log_{10} 0.02}{\log_{10} 0.92} = 46.92 \text{ years.}$$

51. a. An exponential model for the amount of Valium in the bloodstream is $Q = 50 \times \left(\frac{1}{2}\right)^{t/36}$. After 12 hours, there will be $Q = 50 \times \left(\frac{1}{2}\right)^{12/36} = 39.7$ mg.

b. Ten percent of the initial amount is 5 mg, so solve $5 = 50 \times \left(\frac{1}{2}\right)^{t/36}$ for t:

$$0.1 = (0.5)^{t/36} \Rightarrow \log_{10} 0.1 = \log_{10} (0.5)^{t/36} \Rightarrow$$

$$\log_{10} 0.1 = \frac{t}{36} \log_{10} 0.5 \Rightarrow t = \frac{36 \log_{10} 0.1}{\log_{10} 0.5} \Rightarrow$$

$$t = 120 \text{ hours.}$$

53. a. An exponential model for the amount of uranium left is $Q = Q_0 \times \left(\frac{1}{2}\right)^{t/4.5}$, where t is measured in billions of years. We want to determine t when the amount Q is equal to 65% of the original amount, i.e. when $Q = 0.65 Q_0$. Solve $0.65 Q_0 = Q_0 \times \left(\frac{1}{2}\right)^{t/4.5}$ for t – begin by dividing both sides by Q_0:

$$0.65 = \left(\frac{1}{2}\right)^{t/4.5} \Rightarrow \log_{10} 0.65 = \log_{10} (0.5)^{t/4.5} \Rightarrow$$

$$\log_{10} 0.65 = \frac{t}{4.5} \log_{10} 0.5 \Rightarrow t = \frac{4.5 \log_{10} 0.65}{\log_{10} 0.5} \Rightarrow$$

$$t = 2.8 \text{ billion years old.}$$

b. Following the same process shown in part **a**, we have $t = \frac{4.5 \log_{10} 0.45}{\log_{10} 0.5} = 5.2$ billion years old.

FURTHER APPLICATIONS

55. An exponential model for the amount of radioactive substance is $Q = 3 \times \left(\frac{1}{2}\right)^{t/20}$, where Q is the density of the substance in mg/cm^2, and t is time in years, with $t = 0$ corresponding to the present. This model is valid for times in the past – we need only assign negative values for t. Thus 55 years ago corresponds with $t = -55$, so the density at that time was $Q = 3 \times \left(\frac{1}{2}\right)^{-55/20} = 20.2$ mg/cm^2.

One can also solve $3 = Q_0 \times \left(\frac{1}{2}\right)^{55/20}$ for Q_0 to produce the same result.

57. The cart of groceries cost $Q = \$125 \times (1.035)^5 = \$148.46$.

UNIT 10A

QUICK QUIZ

1. **a.** As long as the points are distinct, two points are sufficient to determine a unique line.

3. **a.** By definition, an acute angle has measure less than 90°.

5. **b.** By definition, a right triangle has one 90° angle.

7. **c.** The formula for the volume of a sphere of radius r is $V = \frac{4}{3}\pi r^3$.

9. **c.** The volume of a sphere whose radius has tripled from r to $3r$ is $V = \frac{4}{3}\pi(3r)^3 = 3^3 \cdot \frac{4}{3}\pi r^3$.

DOES IT MAKE SENSE?

21. Does not make sense. If the highways are, in fact, straight lines, they will not intersect more than once.

23. Makes sense. A rectangular prism is just a box.

25. Does not make sense. Basketballs are in the shape of a sphere (a can is shaped like a right circular cylinder).

BASIC SKILLS & CONCEPTS

27. One-half of a circle is 180°.

29. One-tenth of a circle is 36°.

31. One-sixtieth of a circle is 6°.

33. A 45° angle subtends 45°/360° = 1/8 circle.

35. A 120° angle subtends 120°/360° = 1/3 circle.

37. A 270° angle subtends 270°/360° = 3/4 circle.

39. A 300° angle subtends 300°/360° = 5/6 circle.

41. $C = 2\pi \times 8 = 16\pi$ m = 50.3 m. $A = \pi \cdot 8^2 = 64\pi = 201.1$ m^2.

43. $C = 2\pi \times 8 = 16\pi$ cm = 50.3 cm. $A = \pi \cdot 8^2 = 64\pi = 201.1$ cm^2.

45. $C = 2\pi \times 15 = 30\pi$ mm = 94.2 mm. $A = \pi \cdot 15^2 = 225\pi = 706.9$ mm^2.

47. The perimeter is 4 × 9 = 36 mi. The area is 9 × 9 = 81 mi^2.

49. The perimeter is 2 × 12 + 2 × 30 = 84 ft. The area is 30 × 6 = 180 ft^2.

51. The perimeter is 2 × 2.2 + 2 × 2.0 = 8.4 cm. The area is 2.2 × 2.0 = 4.4 cm^2.

53. The perimeter is 6 + 8 + 10 = 24 units. The area is $\frac{1}{2} \times 8 \times 6 = 24$ units^2.

55. The perimeter is 8 + 5 + 5 = 18 units. The area is $\frac{1}{2} \times 8 \times 3 = 12$ units^2.

57. The two semicircular caps together make a single circle, whose radius is 3 feet. The perimeter, then, is 2 × 8 + 2π × 3 = 34.85 ft, and the area is $6 \times 8 + \pi \times 3^2 = 76.27$ ft^2.

59. The area is $\frac{1}{2} \times 12 \times 11 = 66$ ft^2.

61. The area of the parallelogram is $180 \times 150 = 27{,}500$ yd^2.

63. The pool holds $50 \times 30 \times 2.5 = 3750$ m^3.

65. The duct is in the shape of a cylinder, and so its volume is $\pi \times 1.5^2 \times 40 = 282.7$ ft^3 (note that 18 inches is 1.5 feet). The surface area of the duct is $2\pi \times 1.5 \times 40 = 377.0$ ft^2. We are assuming here that there are no circular end pieces on the duct that need to be painted. If there are end pieces, the area would increase by $2\pi \times 1.5^2 = 14.1$ ft^2.

67. The circumference is greater because it is π diameters (of a tennis ball) long, whereas the height is only 3 diameters tall.

69. The volume of the reservoir is $250 \times 60 \times 12 = 180{,}000$ m^3. 30% of this volume, which is 54,000 m^3, must be added.

71. The first tree has greater volume ($\pi \times 2.1^2 \times 40 = 554.2$ ft^3) than the second tree (whose volume is $\pi \times 2.4^2 \times 30 = 542.9$ ft^3).

73. The surface area of the concert hall will be $30^2 = 900$ times as great as the surface area of the model (because areas always scale with the square of the scale factor).

75. The height of the office complex will be 80 times as large as the height of the model.

77. You would need $80^3 = 512{,}000$ times as many marbles to fill the office complex (volumes always scale with the cube of the scale factor).

79. Your waist size is a measurement of length, so it has increased by a factor of 4.

81. Your weight is a function of your volume, so it has increased by a factor of 64.

83. Answers will vary. Example: My waist size is 32 inches, and Sam's is 1.2 × 32 = 38.4 inches.

85. Think of a soup can (a squirrel) and a trash can (a human), compute their surface areas and volumes, find their surface-area-to-volume ratios, and you will discover that the surface-area-to-volume ratio for squirrels is larger than that of humans. Another way to understand this fact: surface area scales with the square of the scale factor, whereas volume scales with the cube of the scale factor. Thus the surface-area-to-volume ratio decreases when objects are scaled up (and in fact, it scales with the reciprocal of the scale factor).

87. The moon's surface-area-to-volume ratio is four (i.e. the scale factor) times as large as Earth's, (see discussion at the end of exercise 85).

89. If you understood the comment at the end of exercise 85, the surface-area-to-volume ratio of the bowling ball should be 1/3 the ratio of the softball, because the bowling ball is 3 times as large. Computing directly: The surface area of the softball is $4\pi \times 2^2 = 50.3$ in^2, its volume is $\frac{4}{3}\pi \times 2^3 = 33.5$ in^3, and its surface-area-to-volume ratio is 1.5. The surface area of the bowling ball is $4\pi \times 6^2 = 452.4$ in^2, its volume is $\frac{4}{3}\pi \times 6^3 = 904.8$ in^3, and its surface-area-to-volume ratio is 0.5. As expected, the ratio of the bowling ball is 1/3 the ratio of the softball.

FURTHER APPLICATIONS

91. a. The book occupies three-dimensional space, so three dimensions are needed to describe it.
    b. The cover of the book occupies a portion of a plane, so two dimensions are needed to describe it.
    c. An edge of the book lies along a line, so a single dimension is adequate for its description.
    d. A point is dimensionless, so anything on the book that corresponds to a point (such as a corner of a page) is of zero dimension.

93. The third line is necessarily perpendicular to both lines due to a theorem from geometry (when two parallel lines are cut by any other line, all three lying in the same plane, alternate interior angles are equal – this amounts to saying that if the angle between one of the parallel lines and the third line is 90°, the angle between the other parallel line and the third line is 90°).

95. a. The surface area of an individual sac is $4\pi \times \left(\frac{1}{6}\right)^2 = 0.349$ mm^2, and the total surface area is 300 million times this value, which is 1.05×10^8 mm$^2 = 105$ m^2. The volume of a single sac is $\frac{4}{3}\pi \times \left(\frac{1}{6}\right)^3 = 0.0194$ mm^3, so the total volume is 5.82×10^6 mm^3 (multiply the individual volume by 300 million).

    b. We need to solve $5.82 \times 10^6 = \frac{4}{3}\pi \times r^3$ for r to find the radius of such a sphere. Multiply both sides of the equation by $\frac{3}{4\pi}$ and take the cube root to find $r = 112$ mm (rounded). This sphere has surface area of $4\pi \times (111.572)^2 = 1.56 \times 10^5$ mm^2. The surface area of the air sacs is

$1.05 \times 10^8 \div 1.56 \times 10^5 = 673$ times as large as the surface area of the hypothetical sphere.

c. We need to solve $1.05 \times 10^8 = 4\pi \times r^2$ for r to find the radius of such a sphere. Divide both sides by 4π and take the square root to find $r = 2891$ mm, or about 2.9 meters (i.e. more than 9 feet). The human lung has a remarkable design as it is able to take advantage of large surface area despite its relatively small volume.

97. The volume of one tunnel is $\frac{1}{2}\pi \times (4)^2 \times 50,000 = 1.26 \times 10^6$ m^3, so the volume of all three is 3.8×10^6 m$^3 = 0.0038$ km^3. Depending on whether the dimensions given are for a rough tunnel, or the structure that a motorist sees driving through it, the amount of earth removed may be considerably more than the answer provided.

UNIT 10B

QUICK QUIZ

1. **c.** Each degree contains 60 minutes.
3. **a.** Lines of latitude run east and west on the globe, parallel to the equator.
5. **b.** Angular size is a function of the distance from the object: when the distance from an object increases, its angular size decreases.
7. **a.** A 10% grade means the slope is 1/10, which implies a vertical change of 1 unit for every horizontal change of 10 units.
9. **a.** The statement in answer **a** is an application of the fact that the ratio of corresponding sides in a pair of similar triangles is equal to the ratio of a different set of corresponding sides.

DOES IT MAKE SENSE?

19. Does not make sense. Points south of the equator experience summer in December.
21. Makes sense. As long as the speaker can ride a bike, a 1% grade should pose no difficulty.
23. Makes sense. Triangle B is just a scaled up version of Triangle A, so they are similar.

BASIC SKILLS & CONCEPTS

25. $32.5° = 32° + 0.5° \times \frac{60'}{1°} = 32°30'$.

27. $12.33° = 12° + 0.33° \times \frac{60'}{1°} = 12°19.8' = 12° + 19' +$
    $0.8' \times \frac{60''}{1'} = 12°19'48''$.

29. $149.83° = 149° + 0.83° \times \dfrac{60'}{1°} = 149°49.8' = 149° + $

$49' + 0.8' \times \dfrac{60''}{1'} = 149°49'48''$.

31. $30°10' = 30° + 10' \times \dfrac{1°}{60'} = 30.17°$.

33. $123°10'36'' = 123° + 10' \times \dfrac{1°}{60'} + 36'' \times \dfrac{1°}{3600''} = $
123.18°.

35. $8°59'10'' = 8° + 59' \times \dfrac{1°}{60'} + 10'' \times \dfrac{1°}{3600''} = 8.99°$.

37. A full circle is 360°, and each degree is 60', so
there are $360° \times \dfrac{60'}{1°} = 21,600'$ in a circle.

39. Copenhagen is at 56°N, 12°E (or 12.5°E).

41. The latitude changes from 36°N to 36°S. To find
the longitude, move 180° east from 115°W to
arrive at 65°E. (Note that after moving 115°, you'll
be at the prime meridian, or 0°, and you still have
65° to go). Thus the point opposite Las Vegas is
36°S, 65°E.

43. Buenos Aires is farther from the North Pole
because its latitude (35°S) is further south than the
latitude of Capetown (34°S).

45. Buffalo is 17° north of Miami, and each degree is
1/360 of the circumference of the Earth. Since the
circumference is about 25,000 mi, Buffalo is
$\dfrac{17}{360} \times 25,000 = 1181$ mi, or about 1200 miles from
Miami.

47. A quarter is about an inch in diameter, and 5 yards
is 180 inches. Thus its angular size is
$1 \text{ in.} \times \dfrac{360°}{2\pi \times 180 \text{ in.}} = 0.32°$.

49. The Sun's true diameter (physical size) is
$0.5° \times \dfrac{2\pi \times 150,000,000 \text{ km}}{360°} = 1.31 \times 10^6 \text{ km}$.

51. A roof with a pitch of 1 in 4 has a slope of 1/4,
which is steeper than a roof with a slope of 2/10.

53. A railroad with a 3% grade has a slope of 3/100 =
0.03, which is not as steep as a railroad with a
slope of 1/25 = 0.04.

55. The slope of a 7 in 12 roof is 7/12. The roof will
rise $\dfrac{7}{12} \times 15 = 8.75$ feet in a horizontal run of 15
feet.

57. Think of a right triangle with side lengths of 6 and
6 – this produces an isosceles triangle, which
implies the non-right angles are both 45°. It is
possible to have a 7 in 6 roof: it's just a steep roof
that rises 7 feet for every 6 feet of horizontal run.

59. The slope of such a road is 20/150 = 0.133, so its
grade is 13.3%.

61. a. Walk 6 blocks east (6/8 mi) and 1 block north
(1/5 mi) for a total distance of 19/20 = 0.95 mi.
b. The Pythagorean theorem gives the straight-line
distance as $\sqrt{(6/8)^2 + (1/5)^2} = 0.78$ mi.

63. a. Walk 2 blocks west (2/8 mi) and 3 blocks north
(3/5 mi) for a total distance of 17/20 = 0.85 mi.
b. The Pythagorean theorem gives the straight-line
distance as $\sqrt{(2/8)^2 + (3/5)^2} = 0.65$ mi.

65. a. Walk 3 blocks east (3/8 mi) and 3 blocks south
(3/5 mi) for a total distance of 39/40 = 0.975 mi.
b. The Pythagorean theorem gives the straight-line
distance as $\sqrt{(3/8)^2 + (3/5)^2} = 0.71$ mi.

67. The height of the triangle in Figure 10.31 is
$\sqrt{800^2 - 150^2} = 785.8$ ft, and thus the area of the lot
is $\dfrac{1}{2} \times 150 \times 785.8 = 58,936 \text{ ft}^2$, which is
$58,936 \text{ ft}^2 \times \dfrac{1 \text{ acre}}{43,560 \text{ ft}^2} = 1.35$ acres.

69. The height of the triangle in Figure 10.31 is
$\sqrt{3800^2 - 600^2} = 3752.3$ ft, and thus the area of the
lot is $\dfrac{1}{2} \times 600 \times 3752.3 = 1,125,700 \text{ ft}^2$, which is
$1,125,700 \text{ ft}^2 \times \dfrac{1 \text{ acre}}{43,560 \text{ ft}^2} = 25.84$ acres.

71. The larger triangle is just a scaled up version of
the smaller triangle (or so it appears), and so they
are similar.

73. The triangle on the left appears to be an isosceles
triangle, whereas the one on the right does not,
which means the triangles don't have the same
angle measure, and cannot be similar.

75. 8/10 = x/5, which implies x = 4. Use the
Pythagorean theorem to show that y = 3 (we know
the triangles are right because the left triangle
satisfies $6^2 + 8^2 = 10^2$).

77. x/60 = 10/40, which implies x = 15. 50/60 = y/40,
which implies y = 100/3 = 33.3.

79. Refer to Figure 10.34, and note that a 12-ft fence
that casts a 25-foot shadow is equivalent to a
house of height h (set back 60 feet from the
property line) that casts an 85-ft shadow. We can
write the ratio 12/25 = h/85, which implies h =
40.8 ft.

81. As in exercise 79, we can write 12/30 = h/80,
which implies h = 32 ft.

83. The radius of a circle with circumference of 60 m
is 60/2π = 9.55 m, and so its area is
$\pi \times (9.55)^2 = 286.5 \text{ m}^2$. The side length of a square
with perimeter of 60 m is 60/4 = 15 m, and so its
area is $15 \times 15 = 225 \text{ m}^2$. The area of the circular
region is larger.

85. The radius of a circle with circumference of 240 m is $240/2\pi = 38.2$ m, and so its area is $\pi \times (38.2)^2 = 4584$ m^2. The side length of a square with perimeter of 240 m is $240/4 = 60$ m, and so its area is $60 \times 60 = 3600$ m^2. The area of the circular region is larger.

87. For the first can described, the area of the top and bottom is $2 \times \pi \times 4^2 = 100.53$ in.^{2}, and the area of the side is $2\pi \times 4 \times 5 = 125.66$ in.^{2}. The cost of the can is 100.53 in.$^2 \times \dfrac{\$1.00}{\text{in.}^2} + 125.66$ in.$^2 \times \dfrac{\$0.50}{\text{in.}^2} = \$163.36$. In a similar manner, it can be shown that the cost of the second can is 157.08 in.$^2 \times \dfrac{\$1.00}{\text{in.}^2} + 125.66$ in.$^2 \times \dfrac{\$0.50}{\text{in.}^2} = \$219.91$.

89. Assuming a rectangular box, the most economical shape is a cube, with side length equal to 2 ft (see Example 10 in the text). Such a box has 6 faces, each with area of 4 ft^2, for a total area of 24 ft^2. At \$0.15 per square foot, the box would cost \$3.60.

FURTHER APPLICATIONS

91. Begin by converting 0.1 second into degrees: $0.1'' \times \dfrac{1'}{60''} \times \dfrac{1°}{60'} = 2.78 \times 10^{-5}$ degree. Now use the angular size formula (solved for physical size) to compute the size of an object whose angular size is 0.1'' at a distance of 350 km. The physical size is $2.78 \times 10^{-5°} \times \dfrac{2\pi \times 350 \text{ km}}{360°} \times \dfrac{1000 \text{ m}}{1 \text{ km}} \times \dfrac{100 \text{ cm}}{1 \text{ m}} = 17$ cm. Thus the letter size would need to be 17 cm (almost 7 inches, much larger than the print in a newspaper) before you could read it with the Hubble, and this neglects the problems of peering through Earth's atmosphere.

93. The throw goes along the hypotenuse of a right triangle, whose length is $\sqrt{90^2 + 90^2} = 127.3$ ft.

95. For simplicity, assume you can row at a rate of 1.0 mph, and you can bike at a rate of 1.5 mph. Using time $= \dfrac{\text{distance}}{\text{rate}}$, the time it takes to bike along the edges of the reservoir is $\dfrac{1.2 \text{ mi} + 0.9 \text{ mi}}{1.5 \text{ mph}} = 1.4$ hr. The time it takes to row along the hypotenuse is $\dfrac{\sqrt{(1.2 \text{ mi})^2 + (0.9 \text{ mi})^2}}{1.0 \text{ mph}} = 1.5$ hr, so biking is faster.

97. a. The volume of the water in the bed is $8 \times 7 \times 0.75 = 42$ ft^3. Divide this volume by the area of the lower room (80 ft^2) to find that the depth of the water is 0.525 ft.

b. The weight of the water in the bed is $42 \text{ ft}^3 \times \dfrac{62.4 \text{ lb}}{1 \text{ ft}^3} = 2620.8$ lb.

99. The perimeter of a corral with dimensions of 10 m by 40 m is 100 m. One could use a square-shaped corral with side length of 20 m to achieve the desired area of 400 square meters; this corral would have a perimeter of 80 m, and would require less fencing.

101. Your project will cost $3 \text{ mi} \times \dfrac{\$500}{1 \text{ mi}} + \sqrt{(2 \text{ mi})^2 + (1 \text{ mi})^2} \times \dfrac{\$1000}{1 \text{ mi}} = \$3736$. Using the plan suggested by your boss will cost $4 \text{ mi} \times \dfrac{\$500}{1 \text{ mi}} + \sqrt{(1 \text{ mi})^2 + (1 \text{ mi})^2} \times \dfrac{\$1000}{1 \text{ mi}} = \$3414$.

103. a. With a lot of work and patience, you might be able to come up with the dimensions of the can that has the lowest surface area (and thus lowest cost): $r = 3.8$ cm and $h = 7.7$ cm. One way to save time is to observe that $355 = \pi r^2 h$, which implies that $h = \dfrac{355}{\pi r^2}$. Insert this expression for h into the surface area formula $2\pi rh + 2\pi r^2$ to find that the surface area can be expressed as a function of r: $A = \dfrac{710}{r} + 2\pi r^2$. Since the cost is directly proportional to the surface area, the object is to find the value of r that yields the smallest surface area A. If you graph A on a calculator, and use the trace function, you'll see that $r = 3.8$ at the low point of the graph.

b. The dimensions of the optimal can found above imply the can is as wide (diameter) as it is high. Soda cans aren't built that way, probably for a variety of reasons. (Among them: the can needs to fit into the hand of an average human, and the top and bottom of the can cost more because the aluminum is thicker).

105. a. Note that the radius of the cone will be 6 ft (because the height is 1/3 the radius). The volume is $\dfrac{1}{3}\pi \times 6^2 \times 2 = 75.4$ ft^3.

b. Again, note that $r = 3h$, and solve $1000 \text{ ft}^3 = \dfrac{1}{3}\pi \times (3h)^2 \times h = 3\pi \times h^3$ for h to find $h = 4.7$ ft.

c. Assuming 10,000 grains per cubic inch, there are $75.4 \text{ ft}^3 \times \left(\dfrac{12 \text{ in.}}{1 \text{ ft}}\right)^3 \times \dfrac{10,000 \text{ grains}}{\text{in.}^3} = 1.3$ billion grains. The figure of 10,000 grains per cubic inch is very much dependent upon the size of sand grain assumed, which varies.

107. a. The pyramid's height is $481 \div 3 = 160.3$ yards, which is about 1.6 times the length of a football field (excluding the end zones).

b. The volume of the pyramid is $\frac{1}{3} \times (756 \text{ ft})^2 \times 481 \text{ ft} = 91,636,272 \text{ ft}^3$, which is

$91,636,272 \text{ ft}^3 \times \left(\frac{1 \text{ yd}}{3 \text{ ft}}\right)^3 = 3,393,936 \text{ yd}^3$.

c. Divide the result in part **b** by 1.5 to get around 2,263,000 blocks.

d.

$2,263,000 \text{ blocks} \times \frac{2.5 \text{ min}}{\text{block}} \times \frac{1 \text{ hr}}{60 \text{ min}} \times \frac{1 \text{ d}}{12 \text{ hr}} \times \frac{1 \text{ yr}}{365 \text{ d}}$

= 21.5 years. This is comparable to the amount of time suggested in historical records, which indicates that Lehner's estimate is in error. However, we don't have all the facts – perhaps Lehner began with the assumption that with a work force of 10,000 laborers, one stone could be placed every 2.5 minutes, or maybe he assumed laborers worked on the project 24 hours per day.

e. The volume of the Eiffel tower is $\frac{1}{3} \times (120 \text{ ft})^2 \times 980 \text{ ft} = 4,704,000 \text{ ft}^3$, which is about 5% of the pyramid's volume.

UNIT 10C

QUICK QUIZ

1. **b**. As noted in the text, fractals successfully replicate natural forms, especially when random iteration is used.

3. **b**. The edge of a leaf (think of a maple leaf) has many of the properties of a fractal, such as self similarity, and a length that increases as one decreases the size of the ruler used to measure it.

5. **c**. The definition of the fractal dimension of an object leads to fractional values for the dimension.

7. **a**. The curve labeled L_6 in Figure 10.53 is simply the sixth step (or iteration) of the process required to generate the snowflake curve. The snowflake curve itself is denoted by L_∞ (the end product of performing infinite iterations).

9. **a**. The definition of a self-similar fractal is that its patterns repeat themselves under greater magnification.

DOES IT MAKE SENSE?

19. Makes sense. The boundary of a rectangle is made of straight lines whose lengths are well defined, and so can be measured with a standard ruler (to a reasonable degree of accuracy).

21. Does not make sense. The boundary of the snowflake island is infinitely long, and though its area is finite, it is not computed with a simple "length times width" formula.

23. Makes sense. Data have been collected that shows many boundaries in nature, such as coastlines and edges of leaves, have fractal dimensions greater than one.

BASIC SKILLS & CONCEPTS

25. According to the definition of fractal dimension, $R = 2$ and $N = 2$, so we have $2 = 2^D$, where D is the fractal dimension. It is evident that $D = 1$ in this case, and thus the object is not a fractal (it behaves like an ordinary geometric object with regard to length measurements).

27. According to the definition of fractal dimension, $R = 2$ and $N = 8$, so we have $8 = 2^D$, where D is the fractal dimension. It is evident that $D = 3$ in this case, and thus the object is not a fractal (it behaves like an ordinary geometric object with regard to volume measurements).

29. According to the definition of fractal dimension, $R = 2$ and $N = 6$, so we have $6 = 2^D$, where D is the fractal dimension. Solving for D using logarithms produces a value of $D = \frac{\log_{10} 6}{\log_{10} 2} = 2.585$, and thus the object is a fractal (it does not behave like an ordinary geometric object with regard to area measurements).

31. According to the definition of fractal dimension, $R = 5$ and $N = 5$, so we have $5 = 5^D$, where D is the fractal dimension. It is evident that $D = 1$ in this case, and thus the object is not a fractal (it behaves like an ordinary geometric object with regard to length measurements).

33. According to the definition of fractal dimension, $R = 5$ and $N = 125$, so we have $125 = 5^D$, where D is the fractal dimension. It is evident that $D = 3$ in this case, and thus the object is not a fractal (it behaves like an ordinary geometric object with regard to volume measurements).

35. According to the definition of fractal dimension, $R = 5$ and $N = 30$, so we have $30 = 5^D$, where D is the fractal dimension. Solving for D using logarithms produces a value of $D = \frac{\log_{10} 30}{\log_{10} 5} = 2.113$, and thus the object is a fractal (it does not behave like an ordinary geometric object with regard to area measurements).

37. a. Suppose that the original line segment has length of 1 unit. When we measure the length of the first iteration of the quadratic Koch curve with a ruler of length 1 unit (or one element), we find the length is 1 unit, the distance from one endpoint to the other. When we shorten the ruler by a factor

of $R = 4$ so that its length is 1/4 unit, the length of the first iteration is $N = 8$ elements (i.e., the length is 8/4 = 2 units). This is true at every stage of the process.

b. According to the definition of fractal dimension, we have $8 = 4^D$. Solving for D yields $D = \dfrac{\log_{10} 8}{\log_{10} 4} = 1.5$. The length of the quadratic Koch curve is infinite, because at each iteration, the length increases by a factor of 2, and thus its length grows without bound as the process continues.

c. Consider what happens to the area of the quadratic Koch island as we go from the first stage (a square) to the second stage. Along the upper boundary, the area is increased by a small square drawn above the boundary, but it is decreased by a small square drawn below the boundary. This is true along every edge of the original square, and the net effect is no change in the area. The same is true for every step of the process, so the area of the final island is equal to the area of the original square. The coastline of the island is infinite, as noted above in part **b**.

FURTHER APPLICATIONS

39. When the ruler is reduced by a factor of $R = 9$, there will be $N = 4$ elements found (each of length 1/9). This gives $4 = 9^D$, which when solved leads to $D = \dfrac{\log_{10} 4}{\log_{10} 9} = 0.631$. (Note that this is the answer you get no matter what stage of the process you choose to analyze). The dimension is less than 1 due to the fact that the end result of the process of constructing the Cantor set is a set of isolated points. Though there are infinitely many points, there aren't "enough" of them to constitute a line segment of measurable length (that is, this fractal object does not behave like an ordinary geometric object with regard to length).

41. a. Note that every time the ruler is decreased in length by a factor of $R = 10$, the number of elements increases by a factor of $N = 20$, which leads to $20 = 10^D$. This implies the fractal dimension is $D = \dfrac{\log_{10} 20}{\log_{10} 10} = 1.301$.

b. Note that every time the ruler is decreased in length by a factor of $R = 2$, the number of area elements increases by a factor of $N = 3$, which leads to $3 = 2^D$. This implies the fractal dimension is $D = \dfrac{\log_{10} 3}{\log_{10} 2} = 1.585$. The fractal dimension is less than two because the standard notion of surface

area for ordinary objects does not carry over to the surface area of fractal objects.

c. Note that every time the ruler is decreased in length by a factor of $R = 2$, the number of volume elements increases by a factor of $N = 6$, which leads to $6 = 2^D$. This implies the fractal dimension is $D = \dfrac{\log_{10} 6}{\log_{10} 2} = 2.585$. A fractal dimension between 2 and 3 is reasonable because such a rock exhibits the same properties as the Sierpinski sponge – it is somewhat "less" than a solid three-dimensional cube because material has been removed in a fractal pattern.

43. The branching in many natural objects has the same pattern repeated on many different scales. This is the process by which self-similar fractals are generated. Euclidean geometry is not equipped to describe the repetitions of patterns on many scales.

UNIT 11A

QUICK QUIZ

1. **b**. Instruments with strings (guitar, violin, piano), woodwinds with reeds (clarinet, bassoon), and instruments with a column of air in a tube (organ pipe, horn, flute) – all produce their sound with an object that vibrates.

3. **a**. Higher pitches go hand in hand with higher frequencies.

5. **a**. Every time you raise the pitch of a sound by an octave, the pitch doubles.

7. **c**. Table 11.1 shows the frequencies of each of the notes in a particular octave; the names of the notes repeat every octave. As you can see in the table, increasing the frequency of middle C by a factor of 1.5 moves you to a G, so if you increase the frequency of the next higher C by a factor of 1.5, you'll end up at the next higher G.

9. **b**. The mathematician Fourier discovered that musical sounds are the sum of several constant-frequency waves.

DOES IT MAKE SENSE?

17. Does not make sense. The pitch of a string is a function of the length of the string, not the number of times the string is plucked.

19. Makes sense. The frequency of each note in the 12-tone scale increases by a factor of $f = \sqrt[12]{2} = 1.05946$, and this is exponential growth.

21. Does not make sense. The scratch on the record would be difficult to repair. However, if it is a minor scratch, one might be able to digitize the sound the record produces, and at that point a digital filter could be used to reduce or remove the effects of the scratch.

BASIC SKILLS & CONCEPTS

23. The next lower octave corresponds to a halving in frequency, so the requested frequencies are 880 cps, 440 cps, 220 cps, and 110 cps.

25.

Note	Frequency (cps)
G	390
G#	413
A	438
A#	464
B	491
C	521
C#	552
D	584
D#	619
E	656
F	695
F#	736

The table can be generated beginning with the initial frequency of 390 cps, and multiplying by $f = \sqrt[12]{2}$ to produce each successive note. Realize that you will get slightly different values for some of the entries in the table if you generate it from middle C, whose frequency is 260 cps. This is due to rounding errors, and it explains the different values shown in this table and in Table 11.1 at the A, A#, and C entries. We began with an initial value of 390 cps to generate this table, but that's a rounded value for the true frequency of middle G. If we had used an exact value for middle G, we should find that the frequency of the C in this table, which is one octave higher than middle C, is 520 (because frequencies double every octave).

27. a. $260 \times \left(\sqrt[12]{2}\right)^{7} = 390$ cps.

b. $260 \times \left(\sqrt[12]{2}\right)^{9} = 437$ cps.

c. An octave is 12 half-steps, so use a total of 19 half-steps. $260 \times \left(\sqrt[12]{2}\right)^{19} = 779$ cps.

d. $260 \times \left(\sqrt[12]{2}\right)^{25} = 1102$ cps.

e. An octave is 12 half-steps, so use a total of 39 half-steps. $260 \times \left(\sqrt[12]{2}\right)^{39} = 2474$ cps.

29. To compute the frequency for a note one half-step below a particular frequency, we simply divide by $f = \sqrt[12]{2}$, which is equivalent to multiplying by the factor f raised to the power of –1. Thus to find the note 7 half steps below middle A, use $437 \times \left(\sqrt[12]{2}\right)^{-7} = 292$ cps. Ten half-steps below middle A is $437 \times \left(\sqrt[12]{2}\right)^{-10} = 245$ cps.

31. The factor by which the tone has increased is $\left(\sqrt[12]{2}\right)^5 = 2^{5/12} = 1.335$. Because $\left(2^{5/12}\right)^{12} = 2^5$, you need 12 fourths. This is the first time the factor of $2^{5/12}$ turns into a factor of 2, and that is what is necessary to return to the same-named note (raising an octave is a doubling in frequency). There are 5 octaves in the circle of fourths, as evidenced by the factor by which the initial frequency has changed: 2^5.

UNIT 11B

QUICK QUIZ

1. **b**. The principal vanishing point in a painting is the point of intersection of those lines in the painting that are parallel in the real scene, and perpendicular to the canvas. We are assuming here that the tracks are, in fact, perpendicular to the canvas, and that they are straight.

3. **a**. Not only does the painting show this fact, but the parallel beams in the ceiling are perpendicular to the canvas, so they should intersect at the principal vanishing point.

5. **c**. If the letter **W** were reflected across this line, it would appear unchanged from the original.

7. **a**. Because a circle can be rotated through any angle and remain unchanged, it has rotation symmetry. It also has reflection symmetry over any line through its center, for the same reason.

9. **a**. Refer to Figure 11.23 in the text.

DOES IT MAKE SENSE?

19. Does not make sense. The principal vanishing point is often very apparent in paintings that are rendered in proper perspective (see Figures 11.6 and 11.8, for example).

21. Makes sense. It is the science and art of perspective painting that allows talented artists to render realistic three-dimensional scenes on a two-dimensional canvas.

23. Does not make sense. Only equilateral triangles, squares, and regular hexagons allow for complete tilings. Susan should rent a tile saw.

BASIC SKILLS & CONCEPTS

25. a. The only obvious vanishing point is depicted in the diagram below. It is not the principal vanishing point because the lines of the road do not meet the canvas at right angles. However, it does lie on the horizon line.

b.

27.

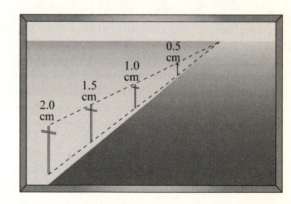

29. a. In order to draw the equally spaced poles, first draw a perspective line along the bases of the existing poles, and draw one along the tops. These two lines should intersect at a vanishing point (directly on, or very near to the horizon). Now measure 2.0 cm along the base line, and draw a vertical pole to meet the top perspective line. Repeat this procedure for the second pole.

b. The heights of the poles can be measured directly with a ruler. You should find the first pole you drew is about 1.0 cm, and the second about 0.5 cm. Notice that the lengths of the poles are decreasing in a linear fashion (2.0, 1.5, 1.0, and 0.5 cm). This is due to the fact that the poles are sandwiched between two lines of constant slope, and that the distances between them are equal. (Also note that the diagram in part **a** is not to scale).

c. If your drawing is an accurate depiction of the scene, the poles are not equally spaced in the actual scene. This is due to perspective: equal distances on the canvas do not correspond with equal distances in the real scene. Because the heights of the poles are assumed to be equal, the short pole you drew in the background is much farther away from you than the first pole you drew, despite the fact that they are equally spaced in the drawing.

31. a. The letters with right/left reflection symmetry are A, H, I, M, O, T, U, V, W, X, and Y.
    b. The letters with top/bottom reflection symmetry are B, C, D, E, H, I, K, O, and X.
    c. The letters with both of these symmetries are H, I, O, and X.
    d. The letters with rotational symmetry are H, I, N, O, S, X, and Z.

33. a. Rotating the triangle about its center 120° or 240° will produce an identical figure.
    b. Rotating the square about its center 90°, 180°, or 270° will produce an identical figure.
    c. Rotating the pentagon around its center 72°, 144°, 216°, or 288° will produce an identical figure.
    d. Rotating an n-gon about its center $360°/n$ (and multiples of this angle less than 360°) will produce an identical figure. There are $n - 1$ different angles of rotational symmetry for an n-gon.

35. The figure has reflection symmetries (about vertical and horizontal lines drawn through its center), and it has rotation symmetry (an angle of 180°).

37. The figure has reflection symmetries (about 6 diagonal lines drawn through its center; these lines go through the spaces between the petals), and it has rotation symmetries (angles of 60°, 120°, 180°, 240°, and 300°).

39.

41.

43.

45. The angles around a point P are precisely the angles that appear inside of a single quadrilateral. Thus, the angles around P have a sum of 360°, and the quadrilaterals around P fit perfectly together.

UNIT 11C

QUICK QUIZ

1. **c.** The value of the golden ratio can be derived by solving $\frac{L}{1} = \frac{L+1}{L}$ for L. This equation comes from the idea that the most aesthetically pleasing way to divide a line is one where the ratio of the length of the long piece (L) to the length of the short piece (1) is equal to the ratio of the length of the entire line ($L + 1$) to the length of the long piece L, and it is this idea that produces the golden ratio.

3. **c.** The value of ϕ is about 1.6, and the ratio of the long piece to the short piece should be around 1.6 before claiming the line is divided into the golden ratio. Since 0.6/0.4 = 1.5, these are roughly the correct lengths for a golden ratio.

5. **b.** A golden rectangle is defined to be a rectangle whose ratio of its long side to short side is ϕ. Note than $10 \div 6.25 = 1.6$.

7. **a.** Refer to Figure 11.46.

9. **b.** The 21st number in the series is the sum of the preceding two numbers.

DOES IT MAKE SENSE?

17. Does not make sense. The ratio of lengths of Maria's two sticks is 2/2 = 1, which is not the golden ratio.

19. Does not make sense. While circles have a lot of symmetry, they do not embody the golden ratio.

BASIC SKILLS & CONCEPTS

21. The longer segment should have a length of 3.71 inches, and the shorter segment a length of 2.29 inches.

23. The middle and right rectangles are golden rectangles.

25. If 5.8 m is the longer side, and S the shorter side, we have $5.8 \div S = \phi$, which means $S = 5.8 \div 1.62 = 3.58$ m. If 5.8 m is the shorter side, and L the longer side, we have $L \div 5.8 = \phi$, which means $L = 5.8 \phi = 9.40$ m.

27. If 0.66 cm is the longer side, and S the shorter side, we have $0.66 \div S = \phi$, which means $S = 0.66 \div 1.62 = 0.41$ cm. If 0.66 cm is the shorter side, and L the longer side, we have $L \div 0.66 = \phi$, which means $L = 0.66 \phi = 1.07$ cm.

FURTHER APPLICATIONS

29. a. Begin with $\dfrac{L}{1} = \dfrac{L+1}{L}$, and multiply both sides by L to arrive at $L^2 = L + 1$, which can be rearranged to produce $L^2 - L - 1 = 0$.

    b. The two roots are
    $$L = \frac{-(-1) + \sqrt{(-1)^2 - 4(1)(-1)}}{2(1)} = \frac{1 + \sqrt{5}}{2} \text{ and}$$
    $$L = \frac{-(-1) - \sqrt{(-1)^2 - 4(1)(-1)}}{2(1)} = \frac{1 - \sqrt{5}}{2}.$$
    The first is $\phi = 1.618034\ldots$

UNIT 12A

QUICK QUIZ

1. **b.** In a three-candidate race, it's possible that none of the candidates receives a majority of the votes.
3. **b.** A 60% vote is required to end a filibuster, which means 41% of the senators (a minority) can prevent a bill from coming to a vote by staging a filibuster.
5. **b.** A preference schedule is a ballot where voters list, in order of their preference, the choices before them.
7. **a.** Look at the second row of Figure 12.3.
9. **b.** Candidate E received only 6 first-place votes.

DOES IT MAKES SENSE?

19. Does not make sense. If both candidates received more than 50% of the vote, the total vote would exceed 100%.
21. Makes sense. Imagine a race with three candidates. Herman could win a plurality of the votes in the first election, with Hanna coming in second, in which case a runoff election would ensue. Hanna would win the election if she beat Herman in the runoff.
23. Makes sense. This is the way most U.S. presidential elections are decided – a candidate wins both the popular and electoral vote.

BASIC SKILLS & CONCEPTS

25. Although Tilden narrowly won the popular vote, Hayes narrowly won the electoral vote, and became president.
27. Garfield narrowly won the popular vote, and easily won the electoral vote to become president.
29. Carter narrowly won the popular vote, and comfortably won the electoral vote to become president.
31. George W. Bush won the popular vote by a close margin, and comfortably won the electoral vote to become president.
33. a. The percentage of shareholders favoring the merger is 6650/10,100 = 65.84%, which is just shy of the required 2/3 = 66.67% vote needed.
    b. A 3/4 vote of a 12-member jury requires at least 9 votes; thus, there will be no conviction.
    c. Only 70% of the states support the amendment (75% is required), so it will fail to pass.
    d. It is likely to become law because 68/100 = 68% of the Senate will support it, and 292/435 = 67.1% of the House will support it, so the veto can be overturned.

35. a. A total of 66 votes were cast.
    b. Candidate D is the plurality winner (22 first-place votes), but not by a majority.
    c. In a runoff, candidate D would win, because 40 voters prefer D to B (who finished in second place in part **b**), while only 26 prefer B to D.
    d. In a sequential runoff, the candidate with the fewest votes is eliminated at each stage. Candidate A received the fewest first-place votes in the original results (only 8), so A is eliminated, and those votes are redistributed among the other candidates. Recounting the votes after A has been eliminated, we find that B has 20 first-place votes, C has 16, and D has 30. Thus C is now eliminated, and the votes counted again. In this final stage, we find that B has 26 first-place votes, and D has 40 votes (as in part **b**), so D wins the sequential runoff.
    e. In a Borda count, first-place votes receive a value of 4, second-place votes receive a value of 3, and so on. Working through each column of the table, and noting the number of voters who cast such ballots, the count for candidate A is computed as follows:
    $$20 \times 1 + 15 \times 3 + 10 \times 2 + 8 \times 4 + 7 \times 3 + 6 \times 3 = 156.$$
    The count for candidate B is
    $$20 \times 4 + 15 \times 1 + 10 \times 1 + 8 \times 1 + 7 \times 2 + 6 \times 2 = 139.$$
    The count for candidate C is
    $$20 \times 2 + 15 \times 2 + 10 \times 4 + 8 \times 2 + 7 \times 1 + 6 \times 4 = 157.$$
    The count for candidate D is
    $$20 \times 3 + 15 \times 4 + 10 \times 3 + 8 \times 3 + 7 \times 4 + 6 \times 1 = 208.$$
    Candidate D wins in this case.
    f. Since there are 4 candidates, there are 6 pairings to consider. The results for each pairing are shown below.

A beats B, 46 to 20.	C beats A, 36 to 30.
D beats A, 52 to 14.	C beats B, 39 to 27.
D beats B, 40 to 26.	D beats C, 50 to 16.

    Since D wins more pairwise comparisons than any other candidate, D is the winner.
    g. As the winner by all five methods, candidate D is clearly the winner of the election.

37. a. A total of 100 votes were cast.
    b. Candidate C is the plurality winner (40 first-place votes), but not by a majority.
    c. In a runoff, candidate A would win, because 55 voters prefer A to C.
    d. With only three candidates, the sequential runoff method is the same as the standard runoff between the top two candidates, and thus A would win.
    e. Using the process illustrated in exercise 35e, the counts of the three candidates are A = 200, B = 210, and C = 190. Candidate B has the highest count, so B wins a Borda count.

f. Since there are 3 candidates, there are 3 pairings to consider. The results of each pairing are:

B beats A, 55 to 45. A beats C, 55 to 45.
B beats C, 55 to 45.

Since B wins two of the pairwise comparisons, B is the winner.

g. Candidate A wins by runoff, B wins by a Borda count and pairwise comparisons, and C is the plurality winner. Thus there is no clear winner.

39. a. A total of 90 votes were cast.

b. Candidate E is the plurality winner (40 first-place votes), but not by a majority.

c. Candidate B wins a runoff, because 50 voters prefer B over E, who would receive 40 votes (D is eliminated at the beginning).

d. With only three candidates earning first-place votes, the sequential runoff method is the same as the standard runoff between the top two candidates, and thus B would win.

e. Using the process illustrated in exercise 35e, the counts of the candidates are A = 260, B = 290, C = 200, D = 290, and E = 310. Candidate E has the highest count, so E wins a Borda cont.

f. Since there are 5 candidates, there are 10 pairings to consider. The results for each pairing are shown below.

A beats B, 60 to 30.   A beats C, 60 to 30.
D beats A, 60 to 30.   E beats A, 70 to 20.
B beats C, 90 to 0.   D beats B, 60 to 30.
B beats E, 50 to 40.   D beats C, 60 to 30.
C beats E, 50 to 40.   E beats D, 70 to 20.

Since D wins more pairwise comparisons than any other candidate (3), D is the winner.

g. Candidates B and E each win two of the five methods, and D wins one method, so all have some claim on the prize. The outcome is debatable.

41. The results of the three head-to-head races are shown below.

A beats B, 18 to 9.   B beats C, 19 to 8.
C beats A, 17 to 10.

No winner can be determined, and the curious situation where voters prefer A to B, B to C, and C to A is called the *Condorcet paradox*, or *voting paradox*.

FURTHER APPLICATIONS

43. a. Fillipo won a plurality, because he won more votes than the other candidates, but no one won a majority, as more than 50% of the vote is required before a majority win can be claimed.

b. Earnest needs just over 23 percentage points to gain a majority and win a runoff. Since 23%/26% = 0.8846, Earnest needs just over 88.46% of Davis's votes.

45. a. King won a plurality because he won more votes than the other candidates, but no one won a majority (King won 382/943 = 40.5% of the vote, which isn't enough).

b. To overtake King, Lord needs 166 additional votes (this would give Lord a total of 472 votes, with King winning only 471 votes).

47.

First	A	A	C	C	D
Second	B	C	A	B	A
Third	C	D	D	A	B
Fourth	D	B	B	D	C
	5	1	2	2	2

49. Each voter will record a first, second, third, fourth, fifth and sixth place vote. The points for these places total 6 + 5 + 4 + 3 + 2 + 1 = 21 points. With 30 voters, there are 30 × 21 = 630 total points.

UNIT 12B

QUICK QUIZ

1. **b**. An election isn't declared fair until all four fairness criteria are satisfied.

3. **b**. Goldsmith won the least number of first-place votes, and so is eliminated in a runoff. Freedman wins Goldsmith's 18 votes, and so beats Berman.

5. **b**. Freedman wins both head-to-head races, and thus by Criterion 2, Freedman should win if this is a fair election.

7. **a**. Candidate X (Freedman) needs to be declared the winner of the first election before Criterion 4 comes into play.

9. **b**. The text gives examples of approval voting scenarios that are not fair (and Arrow's theorem guarantees there are such cases).

DOES IT MAKE SENSE?

15. Makes sense. Karen feels that because she won all head-to-head races, she should win the election so that Criterion 2 is satisfied – otherwise, it's not a fair election.

17. Does not make sense. Assuming the plurality method is used to decide the second election, Table 12.12 claims Criterion 3 is always satisfied (and thus Wendy could not have managed to win).

BASIC SKILLS & CONCEPTS

19. A candidate who wins a majority is the only person to receive a plurality. Thus the candidate will win the election, and Criterion 1 is satisfied.

21. The following preference schedule is just one possible example.

First	B	A	C	C
Second	A	B	A	B
Third	C	C	B	A
	2	4	2	3

C is the plurality winner, but A beats both C and B, which means Criterion 2 is violated.

23. Candidate A would win by the plurality method. However if candidate C were to drop out of the election, then B would win by the plurality method, and Criterion 4 would be violated.

25. Assume a candidate receives a majority. In either runoff method, votes are redistributed as candidates are eliminated. But it is impossible for another candidate to accumulate enough votes to overtake a candidate who already has a majority.

27. By the sequential runoff method (which is the same as the top two runoff method with three candidates), candidate B is eliminated, and C wins the runoff. However, candidate B wins head-to-head races against A and C, so Criterion 2 is violated.

29. In the sequential runoff method, candidate B is eliminated first, then candidate A, making C the winner. Now suppose that the 4 voters on the third ballot (ACB) move C up and vote for the ranking CAB. Now A is eliminated first, and B wins the election. Thus criterion 3 is violated.

31. The following preference schedule is just one possible example.

First	A	B	C
Second	B	A	B
Third	C	C	A
	4	3	5

After candidate B is eliminated, A wins the runoff. However, if C were to drop out, B would win the election. So Criterion 4 is violated.

33. The following preference schedule is just one possible example.

First	A	B	C
Second	B	C	B
Third	C	A	A
	4	2	1

Candidate A has a majority of the votes, but loses in a Borda count (A gets 8 points, B gets 9, and C gets 4, using 2, 1, and 0 points).

35. The following preference schedule is just one possible example.

First	A	D	C
Second	B	B	B
Third	C	A	D
Fourth	D	C	A
	4	8	3

Using the Borda count (with 3, 2, 1, and 0 points), A gets 20 points, B gets 30, C gets 13, and D gets 27 points, making B the winner. However, D wins head-to-head races against all other candidates, and thus Criterion 2 is violated.

37. Using the point system (with 2, 1, and 0 points), candidate A gets 11 points, B gets 11 points, and C gets 14 points, which makes C the winner. However, if Candidate A were to drop from the race, then B would receive 7 points and C would receive 5 points (notice that the point values become 1 and 0 with only two candidates). Thus Criterion 4 is violated.

39. Assume candidate A wins a majority of first-place votes. Then in every head-to-head race with another candidate, A must win (by a majority). Thus, A wins every head-to-head race and is the winner by pairwise comparisons, so Criterion 1 is satisfied.

41. Suppose candidate A wins by the method of pairwise comparisons, and in a second election, moves up above candidate B in at least one ballot. A's position relative to B remains the same or improves, and A's position and B's position relative to the other candidates remains the same, so A must win the second election.

43. The following preference schedule is just one possible example.

First	B	C	D
Second	A	A	C
Third	D	D	B
Fourth	C	B	A
	2	2	2

By the pairwise comparison method, A would beat D, B would beat A, C would beat A and B, and D would beat B and C. This would lead to a tie between C and D. However, if A were to drop out of the election, then D would have two pairwise wins, and C would have one pairwise win. Criterion 4 is violated because the outcome of the election is changed when A drops out.

45. a. Voting only for their first choices, candidate C wins by plurality with 42% of the vote.
b. By an approval vote, 28% + 29% = 57% of the voters approve of A, 28% + 29% +1% = 58%

approve of B, and 42% of the voters approve of C. The winner is B.

47. The electoral votes per person for each of the states can be computed by dividing the number of electoral votes by the population. The results are as follows:

New York 1.61×10^{-6} Illinois 1.65×10^{-6}

Rhode Island 3.70×10^{-6} Alaska 4.58×10^{-6}

It is evident that voters in Alaska have more voting power than those in Illinois.

49. The voters in Rhode Island have more voting power than those in Illinois (see exercise 47).

51. From the least to greatest voting power, the ranking is: New York, Illinois, Rhode Island, Alaska.

FURTHER APPLICATIONS

53. A single runoff would eliminate C and D. Candidate A would win the runoff, 23 to 18. There is no majority winner, so Criterion 1 does not apply. Candidate A beats all other candidates one-on-one, and is also declared winner in a single runoff, so Criterion 2 is satisfied. If A is moved up in any of the rankings, it doesn't affect the outcome of the election, so Criterion 3 is satisfied. If any combination of {B, C, D} drops out of the race, the outcome is not changed. So in this case (though not in general), Criterion 4 is satisfied.

55. Candidate A wins the point system (using 3, 2, 1, and 0 points) with 83 points. There is no majority winner, so Criterion 1 does not apply. Candidate A beats all other candidates one-on-one, and is also declared winner by the point system, so Criterion 2 is satisfied. The point system always satisfies Criterion 3. If any combination of {B, C, D} drops out of the race, the outcome is not changed. So in this case (though not in general), Criterion 4 is satisfied.

57. Candidate A wins by a plurality, but not by a majority. There is no majority winner, so Criterion 1 does not apply. Candidate E beats all other candidates one-on-one, but loses by the plurality method, so Criterion 2 is violated. The plurality method always satisfies Criterion 3. If any of B, C, or D were to drop out of the election, the outcome of A winning would change, so Criterion 4 is violated.

59. The candidates E, D, and B are eliminated sequentially, leaving a final runoff between A and C, which C wins. There is no majority winner, so Criterion 1 does not apply. Candidate E beats all other candidates one-on-one, but loses in the sequential runoff method, so Criterion 2 is violated. If candidate C moved up in any of the rankings, the outcome is not affected, so Criterion

3 is satisfied. If A were to drop out of the election, D would win instead of C, so Criterion 4 is violated.

61. We have seen that E beats all other candidates in pairwise races (see exercise 57) and is the winner by the pairwise comparison method. There is no majority winner, so Criterion 1 does not apply. The pairwise comparison method always satisfies Criteria 2 and 3. It can be shown that if any combination of {A, B, C, D} drops out of the race, the winner is still E, so Criterion 4 is satisfied.

UNIT 12C

QUICK QUIZ

1. **c.** The Constitution does not specify the number of representatives, and the House has, in fact, had 435 members only since 1912. The number briefly rose by two when Hawaii and Alaska were granted statehood (1959), but dropped back to 435 in the next apportionment (and will likely remain at that level, unless a 1941 law is changed).

3. **a.** The standard divisor is defined to be the total U.S. population divided by the number of seats in the House.

5. **b.** In this scenario, the standard divisor would be defined as the population of students divided by the number of teachers, or $25,000 \div 1000 = 25$.

7. **c.** Parks Elementary would get the eight teachers because the Hamilton method assigns the extra teacher to the school with the highest standard quota.

9. **c.** At present, the law states that the Hill-Huntington method is to be used to reapportion seats at every census.

DOES IT MAKE SENSE?

19. Makes sense. If the number of staff support persons needed in a division depends on the number of employees in that division, an apportionment method would be a good idea.

21. Does not make sense. All apportionment methods have deficiencies, and no single method (of those discussed in the text) is better than the others due to the level of math required to carry it out.

BASIC SKILLS & CONCEPTS

23. The number of people per representative would be $350,000,000 \div 435 = 804,598$. If the constitutional limit were observed, the number of representatives would be $350,000,000 \div 30,000 = 11,667$.

25. In order to compute the standard quota, the standard divisor must first be computed. The

standard divisor is $\frac{294 \text{ million}}{435} = 676,000$. For Connecticut, the standard quota is $\frac{3,504,000}{676,000} = 5.18$, which is larger than the 5 seats the state actually has, so Connecticut is underrepresented.

27. The standard divisor is 676,000 (see exercise 25). For Florida, the standard quota is $\frac{17,397,000}{676,000} = 25.74$, which is larger than the 25 seats the state actually has, so Florida is underrepresented.

29. The total number of employees is $250 + 320 + 380 + 400 = 1350$. The standard divisor is then $\frac{1350}{35} = 38.57$. For the first division, the standard quota is $\frac{250}{38.57} = 6.48$. The standard quotas for the other divisions are 8.30, 9.85, and 10.37, respectively (computed in a similar fashion).

31. Refer to exercise 29. Using Hamilton's method for the assignments, round each standard quota down to get $6 + 8 + 9 + 10 = 33$ technicians. Since the first and third divisions have the highest remainders, the extra two technicians will be assigned to them, giving a final apportionment of 7, 8, 10, and 10 technicians to each of the four divisions.

33. The total population is $950 + 670 + 246 = 1866$, so with 100 seats to be apportioned, the standard divisor is $1866 \div 100 = 18.66$. This is used to compute the standard quota and the minimum quotas in the table below. Hamilton's method applied to these three states yields:

State	A	B	C	Total
Pop.	950	670	246	1866
Std. Q.	50.91	35.91	13.18	100
Min. Q.	50	35	13	98
Frac. R.	0.91	0.91	0.18	2
Final A.	51	36	13	100

Assuming 101 delegates, Hamilton's method yields:

State	A	B	C	Total
Pop.	950	670	246	1866
Std. Q.	50.42	35.26	13.32	101
Min. Q.	51	36	13	100
Frac. R.	0.42	0.26	0.32	1
Final A.	52	36	13	101

No state lost seats as a result of the additional available seat, so the Alabama paradox does not occur here.

35. The total population is $770 + 155 + 70 + 673 = 1668$, so with 100 seats to be apportioned, the standard divisor is $1668 \div 100 = 16.68$. This is used to compute the standard quota and the minimum quotas in the table below. Hamilton's method applied to these four states yields:

State	A	B	C	D	Total
Pop.	770	155	70	673	1668
Std. Q.	46.16	9.29	4.20	40.35	100
Min. Q.	46	9	4	40	99
Frac. R.	0.16	0.29	0.20	0.35	1
Final A.	46	9	4	41	100

Assuming 101 delegates, Hamilton's method yields:

State	A	B	C	D	Total
Pop.	770	155	70	673	1668
Std. Q.	46.62	9.39	4.24	40.75	101
Min. Q.	46	9	4	40	99
Frac. R.	0.62	0.39	0.24	0.75	2
Final A.	47	9	4	41	101

No state lost seats as a result of the additional available seat, so the Alabama paradox does not occur here.

37. The total population is $98 + 689 + 212 = 999$, so with 100 seats to be apportioned, the standard divisor is $999 \div 100 = 9.99$. This is used to compute the standard quota and the minimum quotas in the table below. Using a modified divisor of 9.83 instead, we get the modified quotas listed, and the new minimum quotas. Jefferson's method then yields:

State	A	B	C	Total
Pop.	98	689	212	999
Std. Q.	9.81	68.97	21.22	100
Min. Q.	9	68	21	98
Mod. Q.	9.97	70.09	21.57	101.63
N. Min. Q.	9	70	21	100

Since the new minimum quota successfully apportions all 100 seats, we can stop. Note, however, that the quota criterion is violated, because state B's standard quota is 68.97, yet it was given 70 seats.

39. The total population is 979, so with 100 seats to be apportioned, the standard divisor is $979 \div 100 = 9.79$. This is used to compute the standard quota and the minimum quotas in the table below. Using

a modified divisor of 9.60 instead, we get the modified quotas listed, and the new minimum quotas. Jefferson's method then yields:

State	A	B	C	D	Total
Pop.	69	680	155	75	979
Std. Q.	7.05	69.46	15.83	7.66	100
Min. Q.	7	69	15	7	98
Mod. Q.	7.19	70.83	16.15	7.81	101.98
NMQ	7	70	16	7	100

Since the new minimum quota successfully apportions all 100 seats, we can stop. The quota criterion is satisfied.

41. After trial and error, a modified divisor of 38.4 was found to work. The results are summarized in the table below.

Division	I	II	III	IV
Number in division	250	320	380	400
Modified Quota	6.51	8.33	9.90	10.42
Number assigned	7	8	10	10

43. An interesting situation arises in this problem: if the standard divisor, standard quotas, and geometric means are computed without rounding, it turns out a modified divisor is not necessary. That is, after computing said values, the Hill-Huntington method immediately produces an apportionment with no leftover technicians, and it is not necessary to seek a modified divisor. The results are summarized in the table below.

Division	I	II	III	IV
Number in division	250	320	380	400
Standard Quota	6.4815	8.30	9.85	10.37
Geometric Mean	$\sqrt{6\times7}$ 6.4807	$\sqrt{8\times9}$ 8.49	$\sqrt{9\times10}$ 9.49	$\sqrt{10\times11}$ 10.49
Number assigned	7	8	10	10

Division I is the culprit – the standard quota and geometric mean are so close to one another that it is hard to determine whether one should round up or down (note that the values at the other entries in the table are not shown to the same precision, because the decision about whether to round up or down can be made with less precise values). If the rounded values from exercise 30 are used, you will need to use a modified divisor in order to find an apportionment that uses all 35 technicians. A

divisor of 38.4 is suitable (and realize that modified divisors are not unique, in that divisors close to one another produce the same results).

FURTHER APPLICATIONS

45. The total population is 7710, so with 100 seats to be apportioned, the standard divisor is $7710 \div 100 = 77.10$. This is used to compute the standard quota and the minimum quotas in the table below. Hamilton's method applied to these three states yields:

State	A	B	C	Total
Pop.	1140	6320	250	7710
Std. Q.	14.79	81.97	3.24	100
Min. Q.	14	81	3	98
Frac. R.	0.79	0.97	0.24	2
Final A.	15	82	3	100

With the addition of a new state D of population 500, for whom 5 new delegates are added, Hamilton's method (with a new standard divisor of $8210 \div 105 = 78.19$) yields:

State	A	B	C	D	Total
Pop.	1140	6320	250	500	8210
Std. Q.	14.58	80.83	3.20	6.39	105
Min. Q.	14	80	3	6	103
Frac. R.	0.58	0.83	0.20	0.39	2
Final A.	15	81	3	6	105

Even though 5 new seats were added with state D representation in mind, Hamilton's method assigned 6 seats to this state, 1 of them at the expense of state B. Since B lost a seat as a result of the additional seats for the new state, the New State paradox occurs here.

47. a. The total population is 999, so with 100 seats to be apportioned, the standard divisor is $999 \div 100 = 99.9$. This is used to compute the standard quota and the minimum quotas in the table below. Hamilton's method yields:

State	A	B	C	Total
Pop.	535	334	120	999
Std. Q.	53.55	34.43	12.01	100
Min. Q.	53	34	12	99
Frac. R.	0.55	0.43	0.01	1
Final A.	54	34	12	100

b. Jefferson's method begins as with Hamilton's. As noted in part **a**, the standard divisor is 9.99, so we try lower modified divisors until the apportionment comes out just right. By trial and error, the modified divisor 9.90 is found to work, as documented in the last two rows of the table below. Note that this choice of the modified divisor is not unique, as other nearby values also work.

State	A	B	C	Total
Pop.	535	334	120	999
Std. Q.	53.55	34.43	12.01	100
Min. Q.	53	34	12	98
Mod. Q.	54.04	34.75	12.12	100.91
N. Min. Q.	54	34	12	100

c. Webster's method requires us to find a modified divisor such that the corresponding modified quotas round (not truncate) to numbers that sum to the desired 100 delegates. Inspecting the table in part **a**, we see that the standard divisor and standard quotas are already adequate – we don't need to seek a modified divisor in this case.

State	A	B	C	Total
Pop.	535	334	120	999
Std. Q.	53.55	34.43	12.01	100
Min. Q.	53	34	12	99
Rou. Q.	54	34	12	100

d. The Hill-Huntington method requires us to find a modified divisor such that the corresponding modified quotas, *rounded relative to the geometric mean*, yield numbers which sum to the required 100 delegates. As in part **c**, the standard divisor and standard quotas are already sufficient because they round (per the geometric means) to the desired apportionment. Thus there is no need to seek out a modified divisor in this case.

State	A	B	C	Total
Pop.	535	334	120	999
Std. Q.	53.55	34.43	12.01	100
Min. Q.	53	34	12	99
Geo. M.	53.50	34.50	12.49	-
Rou. Q.	54	34	12	100

The geometric means are for the whole numbers bracketing the standard quotas, namely: $\sqrt{53 \times 54}$ = 53.50, $\sqrt{34 \times 35}$ = 34.50, and $\sqrt{12 \times 13}$ = 12.49. The modified quotas (in this case the standard quotas) are compared to these, and hence 34.43 and 12.01 are rounded down to 34 and 12, respectively, whereas 53.55 is rounded up to 54.
e. All four methods gave the same results.

49. Refer to exercise 47 for a detailed explanation of the method of solution that yields the following apportionments.
a. Standard divisor = 100.

State	A	B	C	D	Total
Pop.	836	2703	2626	3835	10,000
Std. Q.	8.36	27.03	26.26	38.35	100
Min. Q.	8	27	26	38	99
Frac. R.	0.36	0.03	0.26	0.35	1
Final A.	9	27	26	38	100

b. Modified divisor = 98.3.

State	A	B	C	D	Total
Pop.	836	2703	2626	3835	10,000
Std. Q.	8.36	27.03	26.26	38.35	100
Min. Q.	8	27	26	38	99
Mod. Q.	8.50	27.50	26.71	39.01	101.73
NMQ	8	27	26	39	100

c. Modified divisor = 99.5.

State	A	B	C	D	Total
Pop.	836	2703	2626	3835	10,000
Std. Q.	8.36	27.03	26.26	38.35	100
Min. Q.	8	27	26	38	99
Mod. Q.	8.40	27.17	26.39	38.54	100.50
Rou. Q.	8	27	26	39	100

d. Modified divisor = 99.5.

State	A	B	C	D	Total
Pop.	836	2703	2626	3835	10,000
Std. Q.	8.36	27.03	26.26	38.35	100
Min. Q.	8	27	26	38	99
Mod. Q.	8.40	27.17	26.39	38.54	100.49
Geo. M.	8.49	27.50	26.50	38.50	-
Rou. Q.	8	27	26	39	100

e. The Jefferson, Webster, and Hill-Huntington methods all gave the same result, so one could argue that these yield the best apportionment.

51. a. The total population is 390 students, so with 10 committee positions to be apportioned, the standard divisor is 390 ÷ 10 = 39. This is used to compute the standard quota and the minimum quotas in the table below. Hamilton's method yields:

Group	Soc.	Pol.	Ath.	Total
Pop.	48	97	245	390
Std. Q.	1.23	2.49	6.28	10
Min. Q.	1	2	6	9
Frac. R.	0.23	0.49	0.28	1
Final A.	1	3	6	10

b. Jefferson's method begins as with Hamilton's. As noted in part **a**, the standard divisor is 39, so we try lower modified divisors until the apportionment comes out just right. By trial and error, the modified divisor of 35 is found to work. Note that this choice of the modified divisor is not unique, as other nearby values also work.

Group	Soc.	Pol.	Ath.	Total
Pop.	48	97	245	390
Std. Q.	1.23	2.49	6.28	10
Min. Q.	1	2	6	9
Mod. Q.	1.37	2.77	7.00	11.14
NMQ	1	2	7	10

c. Webster's method requires us to find a modified divisor such that the corresponding modified quotas round (not truncate) to numbers that sum to the desired 10 members. Inspecting the tables in parts **a** and **b**, we see that neither the standard nor modified quotas there work, so we must try other modified divisors. This time, we find that 38 is a suitable modified divisor. Note that this choice of the modified divisor is not unique, as other nearby values also work.

Group	Soc.	Pol.	Ath.	Total
Pop.	48	97	245	390
Std. Q.	1.23	2.49	6.28	10
Min. Q.	1	2	6	9
Mod. Q.	1.26	2.55	6.45	10.26
Rou. Q.	1	3	6	10

d. The Hill-Huntington method requires us to find a modified divisor such that the corresponding modified quotas, *rounded relative to the geometric mean*, yield numbers that sum to the desired 10 members. The standard divisor of 39 and standard quotas are already sufficient because they round (per the geometric means) to the desired apportionment. Thus there is no need to seek out a modified divisor in this case.

Group	Soc.	Pol.	Ath.	Total
Pop.	48	97	245	390
Std. Q.	1.23	2.49	6.28	10
Min. Q.	1	2	6	9
Geo. M.	1.41	2.45	6.48	-
Rou. Q.	1	3	6	10

The geometric means are for the whole numbers bracketing the standard quotas, namely: $\sqrt{1 \times 2} = 1.41$, $\sqrt{2 \times 3} = 2.45$, and $\sqrt{6 \times 7} = 6.48$. The modified quotas (in this case the standard quotas) are compared to these, and hence 1.23 and 6.28 are rounded down to 1 and 6, respectively, whereas 2.49 is rounded up to 3.

e. The Hamilton, Webster, and Hill-Huntington methods all give the same result, and thus one could argue that these yield the best apportionment.

53. Refer to exercise 51 for a detailed explanation of the method of solution that yields the following apportionments.
a. Note that the total "population" is 2.5 + 7.6 + 3.9 + 5.5 = 19.5 (million dollars). Standard divisor = 0.78.

Store	Bou.	Den.	Bro.	F. C.	Total
Pop.	2.5	7.6	3.9	5.5	19.5
Std. Q.	3.21	9.74	5.00	7.05	25
Min. Q.	3	9	5	7	24
Frac. R.	0.21	0.74	0.00	0.05	1
Final A.	3	10	5	7	25

b. Modified divisor = 0.76.

Store	Bou.	Den.	Bro.	F. C.	Total
Pop.	2.5	7.6	3.9	5.5	19.5
Std. Q.	3.21	9.74	5.00	7.05	25
Min. Q.	3	9	5	7	24
Mod. Q.	3.29	10.00	5.13	7.24	25.66
NMQ	3	10	5	7	25

c. Modified divisor not necessary.

Store	Bou.	Den.	Bro.	F. C.	Total
Pop.	2.5	7.6	3.9	5.5	19.5
Std. Q.	3.21	9.74	5.00	7.05	25
Min. Q.	3	9	5	7	24
Rou. Q.	3	10	5	7	25

d. Modified divisor not necessary.

Store	Bou.	Den.	Bro.	F. C.	Total
Pop.	2.5	7.6	3.9	5.5	19.5
Std. Q.	3.21	9.74	5.00	7.05	25
Min. Q.	3	9	5	7	24
Geo. M.	3.46	9.49	5.48	7.48	-
Rou. Q.	3	10	5	7	25

e. All four methods give the same results.

UNIT 12D

QUICK QUIZ

1. **c.** The process of redrawing district boundaries is called redistricting.

3. **b.** The number of Republican and Democratic voters is close to the same in most states, so districts where there was a large margin of victory are likely to be districts where there is a heavy concentration of Republican voters.

5. **b.** If district boundaries were drawn in a random fashion, one would expect that Republicans would win about 58% of the house seats. Since they won

21/32 = 66% of the seats, it appears that the district boundaries were set in a way that favored Republicans.

7. **a.** If you concentrate most of the Republicans in a few districts (and essentially concede that they will win in those districts), you will stack the deck in the favor of the Democrats in all the remaining districts.

9. **a.** If district lines are drawn in such a way as to maximize the number of, say, Democratic seats in the House, they must concentrate the Republican voters in a few districts, and these districts are going to elect Republican representatives (assuming voters vote along party lines).

DOES IT MAKE SENSE?

17. Makes sense. The district lines in that state are most likely drawn in such a way as to favor Democrats.

19. Does not make sense. If the practice of gerrymandering is alive and well in your state, you probably shouldn't expect that the percentage of voters from the various political parties is accurately reflected in the percentage of representatives from those parties that win elections.

21. Does not make sense. Current laws require that district lines should be drawn to produce contiguous districts.

BASIC SKILLS & CONCEPTS

23. **a.** In 1998, the percentage of votes cast for Republicans was $\frac{574}{574+407}=59\%$, and for Democrats it was $\frac{407}{574+407}=41\%$. In 2004, the percentage of votes cast for Republicans was $\frac{1126}{1126+598}=65\%$, and for Democrats it was $\frac{598}{1126+598}=35\%$.

b. In 1998, the percentage of House seats won by Republicans was 5/6 = 83%, and for Democrats it was 1/6 = 17%. In 2004, the percentage of House seats won by Republicans was 6/8 = 75%, and it was 2/8 = 25% for Democrats.

c. In both 1998 and 2004, the percentage of House seats won by Republicans was significantly larger than the percentage of votes they received, though the disparity was not as pronounced in 2004.

d. Because of the change noted in part **c**, it's plausible that redistricting had an effect on the distribution of representatives.

25. **a.** In 1998, the percentage of votes cast for Republicans was $\frac{1040}{1040+592}=64\%$, and for Democrats it was $\frac{592}{1040+592}=36\%$. In 2004, the percentage of votes cast for Republicans was $\frac{1914}{1914+1366}=58\%$, and for Democrats it was $\frac{1366}{1914+1366}=42\%$.

b. In 1998, the percentage of House seats won by Republicans was 8/11 = 73%, and for Democrats it was 3/11 = 27%. In 2004, the percentage of House seats won by Republicans was 7/13 = 54%, and it was 6/13 = 46% for Democrats.

c. In 1998, the percentage of House seats won by Republicans was significantly larger than the percentage of votes they received. In 2004, the percentage of votes and seats were pretty close to one another.

d. Because of the change noted in part **c**, it's plausible that redistricting had an effect on the distribution of representatives.

27. **a.** In 1998, the percentage of votes cast for Republicans was $\frac{1472}{1472+1381}=52\%$, and for Democrats it was $\frac{1381}{1472+1381}=48\%$. In 2004, the percentage of votes cast for Republicans was $\frac{2565}{2565+2478}=51\%$, and for Democrats it was $\frac{2478}{2565+2478}=49\%$.

b. In 1998, the percentage of House seats won by Republicans was 10/21 = 48%, and for Democrats it was 11/21 = 52%. In 2004, the percentage of House seats won by Republicans was 12/19 = 63%, and it was 7/19 = 37% for Democrats.

c. In 1998, the percentage of House seats won by Republicans was very close to the percentage of votes they received. In 2004, the percentage of votes changed only by a small amount, yet the percentage of seats won went up by a significant margin.

d. Because of the change noted in part **c**, it's plausible that redistricting had an effect on the distribution of representatives.

29. **a.** The most likely distribution would be 8 Republican and 8 Democrat seats.

b. The maximum number of Republican seats that could be won is 15. In order to get a majority of Republican voters in as many districts as possible, you need to load up one district with Democrats. Each district has 625,000 people – imagine drawing district lines so that all 625,000 people were Democrats in a particular district. That district would elect a Democrat for its representative, but the remaining 15 districts could be drawn to have Republican majorities, and they would elect Republicans. (Note: With the assumption that voting is to take place along party lines, all you'd really need to do is move 1 Democrat from each of 15 districts into the remaining district, and then take 15 Republicans from that district and distribute them evenly across the other 15 districts. This would produce a Republican majority in 15 districts. Of course this assumes that each district begins with exactly 312,500 Republicans, and 312,500 Democrats). Reverse the above logic to convince yourself that the minimum number of Republican representatives is 1.

31. a. The most likely distribution would be 6 Republican and 6 Democrat seats.

b. The maximum number of Republican seats is 11, and the minimum number is 1. (Refer to exercise 29b; the logic is the same).

33. a. The most likely distribution would be 15 Republican and 0 Democrat seats. This is due to the fact that with random district lines, one would expect an 80% majority in favor of Republicans in every single district, and as long as voting takes place along party lines, all of these districts will elect Republican representatives.

b. From part **a**, we see that the maximum number of Republican seats is 15. In order to find the minimum number, first observe that each district has 500,000 people, and the entire state has 1,500,000 Democrats. A particular district will elect a Democrat only when the district has a Democrat majority. This requires at least 250,001 Democrats (and 249,999 Republicans), with the assumption that voting takes place along party lines (and that everyone votes). However, there are only 1,500,000 Democrats to spread around, and since 1,500,000 ÷ 250,001 = 5.999976, the maximum number of districts where one could find a Democrat majority is 5. This implies that the minimum number of Republican seats is 10.

35. There are several solutions. Perhaps the easiest to describe is one where the boundaries divide the state into 8 rectangles, each 4 blocks wide and 2 blocks high.

37. There are several solutions. Perhaps the easiest to describe is one where the boundaries divide the state into 8 rectangles, each 4 blocks wide and 2 blocks high.

39. It is not possible to draw district boundaries that satisfy the conditions given. Three Democrats are needed in a single district for a Democrat majority, and thus a valid solution would require three Democrats in each of three districts, for a total of nine Democrats. But there are only eight Democrats in the state.

41. Answers will vary (there are several solutions).

FURTHER APPLICATIONS

43. Answers will vary for the first three cases (there are several solutions). The last case, where 4 Republicans are elected, is not possible. Five Republicans are needed in a single district for a Republican majority, and thus a valid solution would require five Republicans in each of four districts, for a total of 20 Republicans. But there are only 18 Republicans in the state.

45. There are many solutions.

47. It is not possible for one party to win every House seat for the situation described. To show this for the example provided (20 voters and four districts), note that 3 voters of the same party are required for a majority in a single district. Thus a majority in all four districts would require 12 voters of the same party, which is impossible, because there are only 10 such voters. This argument can be generalized to give a proof of the statement in the exercise.

UNIT 13A

QUICK QUIZ

1. **c.** In going from the East Island to the south shore, one crosses the bridge (or edge) labeled *e*.

3. **b.** The edges in the network represent borders with other countries.

5. **b.** An Euler circuit exists for a network if and only if each vertex has an even number of edges.

7. **b.** A tree contains no circuits.

9. **a.** This is not a spanning network because it does not include the vertex *C*.

DOES IT MAKE SENSE?

21. Makes sense. A network could be used to represent those countries (vertices) that trade with other countries.

23. Does not make sense. An Euler circuit guarantees the salesman will visit the town in which he begins at least twice.

25. Does not make sense. A minimum cost spanning network may not give an optimal walking route.

BASIC SKILLS & CONCEPTS

27. a.

b. The network would look the same as in part **a**, except that every edge would be replaced by two edges (each edge representing a sidewalk).

29.

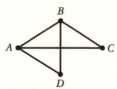

Rather than thinking of Amy, Beth, Cate, and Daniel and the baseball cards they trade, think of various countries and the commodities that they trade.

31. a. Four vertices are needed to represent the land masses, and five edges are needed to represent the ferry routes.

b.

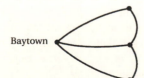

c. No, it is not possible, as the path described would be an Euler circuit. But no Euler circuit exists for this network, because two of the vertices have an odd number of edges.

33. Intersections are represented by vertices, and sidewalks by edges in the network pictured below.

Because every vertex has an even degree, an Euler circuit can be found (there are, in fact, many in this network).

35. a. There are eight vertices, so the network is of order 8.

b. Vertices *A*, *D*, *E*, *F*, *G*, and *H* have degree 3; vertices *B* and *C* have degree 4.

c. The network has no special form.

d. With vertices of odd degree, there are no Euler circuits.

37. a. There are eight vertices, so the network is of order 8.

b. All vertices have degree 4.

c. The network has no special form that was mentioned in the text (it's called a *bipartite* since every vertex of the top set of four vertices is joined to every vertex of the bottom set of four vertices).

d. Because all vertices have even degree, there are Euler circuits.

e. One Euler circuit is *A*, *E*, *B*, *F*, *C*, *G*, *D*, *H*, *C*, *E*, *D*, *F*, *A*, *G*, *B*, *H*, *A*.

39. a. There are five vertices, so the network is of order 5.

b. Vertex *A* has degree 1; vertices *C* and *D* have degree 2; vertex *E* has degree 3; vertex *B* has degree 4.

c. The network has no special form.

d. With vertices of odd degree, there are no Euler circuits.

41. To carry out Kruskal's algorithm, highlight edges of lowest cost to highest cost until a spanning network has been found. In those cases where a closed circuit exists, delete edges of highest cost. Following the algorithm for this circuit results in edges *AB*, *BC*, *BE*, *EF*, and *BD*, with a cost of 14. Note that the edge *BD* could be replaced by *ED*, and you'd still have a minimum cost spanning network (these networks are not always unique).

43. To carry out Kruskal's algorithm, highlight edges of lowest cost to highest cost until a spanning network has been found. In those cases where a closed circuit exists, delete edges of highest cost. Following the algorithm for this circuit results in edges *d*, *b*, *f*, *h*, *a*, and *c*, with a length of 13 km.

FURTHER APPLICATIONS

45. Answers may vary.

a.

b.

c.

47. The complete network of order 5 is shown below.

The network has 10 edges, which agrees with the rule

$$\frac{n \times (n-1)}{2} = \frac{5 \times (5-1)}{2} = 10 \ .$$

49. The order of the network is 9, because 8 states border Tennessee, and for the same reason, the degree of the Tennessee vertex is 8.

51. a.

b.

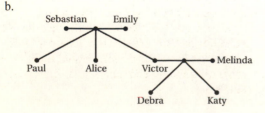

53.

–	A	B	C	D	E	F
A	–	X	X	X	X	
B	X	–	X		X	X
C	X	X	–	X		X
D	X		X	–	X	X
E	X	X		X	–	X
F		X	X	X	X	–

UNIT 13B

QUICK QUIZ

1. **c.** A Hamiltonian circuit is a closed path through a network that passes through every vertex exactly once, and returns to the starting vertex.

3. **b.** In a complete network, every vertex is connected to every other vertex.

5. **b.** There are $\dfrac{(6-1)!}{2} = 5 \times 4 \times 3$ Hamiltonian circuits in a complete network of order 6.

7. **c.** In general, there is no known algorithm that produces the optimal solution to a traveling salesman problem, other than checking every possible Hamiltonian circuit.

9. **a.** A near-optimal solution is one where we don't check every conceivable circuit, but we use an algorithm (such as the nearest neighbor method) to produce an answer that is close to the optimal solution.

DOES IT MAKE SENSE?

15. Does not make sense. Justin might want to visit cousin Jed, return to his house, visit cousins Jean and Jorge, and once again return to his house, followed by visits to his other cousins, in which case the traveling salesman problem would not apply (because Justin would be traveling through the vertex of his house several times). There's nothing in the statement of the problem that clearly delineates the need for a traveling salesman approach.

17. Makes sense. Roger need only test $\dfrac{(5-1)!}{2} = 12 \text{ circuits}$, which would be reasonable task to attempt in an evening.

BASIC SKILLS & CONCEPTS

19. The circuit shown is not a Hamiltonian circuit because it doesn't return to the starting point. It can be made into a Hamiltonian circuit by rotating the arrow along the vertical edge so that it points to the leftmost vertex.

21. The circuit shown is not a Hamiltonian circuit because it doesn't return to the starting point. It cannot be made into a Hamiltonian circuit.

23. a. There are $\dfrac{(7-1)!}{2} = 360$ Hamiltonian circuits in a complete network of order 7.

    b. There are $\dfrac{(15-1)!}{2} = 4.36 \times 10^{10}$ Hamiltonian circuits in a complete network of order 15.

    c. It would take $360 \text{ s} \times \dfrac{1 \text{ min}}{60 \text{ s}} = 6$ minutes to check the circuits in part **a**. For part **b**, it would take about $4.36 \times 10^{10} \text{ s} \times \dfrac{1 \text{ hr}}{3600 \text{ s}} \times \dfrac{1 \text{ d}}{24 \text{ hr}} \times \dfrac{1 \text{ yr}}{365 \text{ d}} = 1383$ years. (Note that this answer uses the rounded value of 4.36×10^{10} circuits – using a more exact number would produce a rounded answer of 1382 years).

25. a. The network will look exactly like Figure 13.35b of the text except that the vertex for Zion would be missing, along with all of the edges connected to it. The resulting network would have four vertices, and a total of six edges.

    b. The length of the trip is $136 + 75 + 151 + 108 = 470$ miles.

    c. Travel from Grand Canyon to Bryce to Capitol Reef to Canyonlands, and return to Grand Canyon, for a total distance of $108 + 69 + 75 + 202 = 454$ miles.

    d. Travel from Bryce to Capitol Reef to Canyonlands to Grand Canyon, and return to Bryce for a total distance of $69 + 75 + 202 + 108 = 454$ miles.

    e. There are three different Hamiltonian circuits possible in this network of order four. They have lengths of 454, 470, and 558 miles, so we have found the Hamiltonian circuit with the shortest length in parts **c** and **d**.

27. a. There are $(12 - 1)! \div 2 = 19{,}958{,}400$ circuits.

    b. It will take $19{,}958{,}400 \text{ circuits} \times \dfrac{1 \text{ s}}{1{,}000{,}000 \text{ circuits}}$
    $= 19.96$, or about 20 seconds.

    c. That's about the time needed to take two breaths.

29. a. There are $\dfrac{(20-1)!}{2} = 6.08 \times 10^{16}$ circuits.

    b. It will take $6.08 \times 10^{16} \text{ circuits} \times \dfrac{1 \text{ s}}{1{,}000{,}000 \text{ circuits}}$
    $= $ about 60 billion seconds, or about 1930 years.

    c. That's almost two millennia.

31. Note that because this network is not complete, it is difficult to determine exactly how many Hamiltonian circuits exist, unless you spend the time to count them all. Also true is that the nearest neighbor method does not always work for networks that aren't complete (see below). The six circuits that the exercise asks us to check – those beginning at each city, and using the nearest neighbor method – follow here.

    Circuit *AEFBCDX* fails to produce a complete circuit when the nearest neighbor method is applied. All goes well until we get to vertex *D*, at which point we cannot return to *A* (because the network is not complete).

    The circuit beginning at city *B* also fails to produce a Hamiltonian circuit. Using the nearest neighbor method, we travel along the path *BFEAX*, at which point we are stuck, because we cannot go from *A* to a vertex not yet visited.

    Beginning at city *C*, we follow the path *CDF*, at which point we have a decision to make: it is 2 hours to both cities *E* and *B*. Choosing the first option leads us to the first complete circuit available (again, using the nearest neighbor method): *CDFEABC*, which would take 23 hours to traverse. The other option (*CDFBEAX*) runs into a dead end at vertex *A*, because we cannot return to *C* from there.

    The circuit beginning at *D* is just a repeat of the circuit beginning at *C*. We come to a decision point at *F*; one way produces the circuit *DFEABCD* (which has length 23, and is the same circuit as described above, with a different initial point), and the other way produces the dead end circuit *DFBEAX*.

    Starting at *E*, we run into a dead end: *EFBAX* (this is similar to what happened with the circuit beginning at *B*).

    Finally, the circuit starting at *F* gives us *FEABCDF*, a circuit of length 23, and identical to the Hamiltonian circuits beginning at *C* and *D* (with different initial point). The nearest neighbor method also produces the dead end circuit *FBEAX*. If you break from convention and don't follow the nearest neighbor rule, you can find circuits that are as good as those found above (*ABFCDEA*, with length 23, for example).

FURTHER APPLICATIONS

33. There are three possible Hamiltonian circuits in this complete network of order 4. These are *ACBDA*, with length 15; *ACDBA*, with length 14; and *ABCDA*, with length 17. Abe should, of course, use the second route to minimize travel distance, though the nearest neighbor method produces the first route (recall that the nearest neighbor method usually produces a near optimal solution). The other drivers should use the second route, beginning and ending at their respective houses.

35. The company will fly for 30 fewer hours each day, for a savings of $15,000 \times 30 = \$450,000$ per day. Multiply this by 365 to find a savings of $164,250,000 per year.

UNIT 13C

QUICK QUIZ

1. **b**. One objective in using CPM (critical path method) to solve scheduling problems is to find the most efficient way to carry out a plan.

3. **c**. When two (or more) tasks can occur at the same time between two stages of the project, the task that requires the most time is called the limiting task.

5. **b**. The amount of slack time for *deliver construction materials* is 1 month, because it takes two months to complete the *finance* task.

7. **a**. Changing *design* to 1.5 months would save 1.5 months, whereas changing *construction* to 2 months would save only 1 month as the other simultaneous tasks between those vertices take 3 months. Changing *finance* would also save only 1 month.

9. **a**. The EFT for any task is its EST added to the amount of time it takes to complete the task (and in this case, that amounts to 7 months).

DOES IT MAKE SENSE?

17. Does not make sense. Vacations are usually sequential affairs in which one stop/city occurs at a time. The critical path method applies to problems that require concurrent tasks.

19. Does not make sense. A delay in tasks that are included in the critical path always delays the entire project.

BASIC SKILLS & CONCEPTS

21. Tasks a, e, f, and g are on the critical path between A and F because these take the longest time.

23. Tasks a, e, f, g, i, and h are on the critical path for the entire project because these take the longest time.

25. Tasks a, d, e, and g are on the critical path between A and F because these take the longest time.

27. Tasks a, d, e, g, and h are on the critical path for the entire project because these take the longest time.

29. a. The EST for task e is 2 hours, and for i it is 7.5 hours.
    b. The LST for task e is 2 hours, and for i it is 7.5 hours.

31. a. The EFT for task e is 2.5 hours, and for i it is 9.5 hours.
    b. The LFT for task e is 2.5 hours, and for i it is 9.5 hours.

33. a. The EST for task e is 4 months, and for h it is 15 months.
    b. The LST for task e is 4 months, and for h it is 15 months.

35. a. The EFT for task e is 7 months, and for h it is 19 months.
    b. The LFT for task e is 7 months, and for h it is 19 months.

37. Since task h is not on the critical path, and since it has a slack time of 1 hour, there is no effect on the overall schedule if it is delayed by 1 hour.

39.

Task	EST	LST	EFT	LFT	Slack
a	0.0	0.0	2.0	2.0	0.0
b	0.0	1.0	1.0	2.0	1.0
c	0.0	1.5	1.0	2.5	1.5
d	1.0	2.0	1.5	2.5	1.0
e	2.0	2.0	2.5	2.5	0.0
f	2.5	2.5	4.5	4.5	0.0
g	4.5	4.5	7.5	7.5	0.0
h	4.5	5.5	6.5	7.5	1.0
i	7.5	7.5	9.5	9.5	0.0
j	9.5	10.0	10.0	10.5	0.5
k	9.5	9.5	10.5	10.5	0.0

41. Since task e is on the critical path, a one-month delay in this task will delay the entire project by one month.

43. The slack time for a task is LST – EST (or, equivalently, LFT – EFT). For this job, the slack times are $a = 0$; $b = 0$; $c = 10$; $d = 0$; $e = 15$; $f = 30$; $g = 0$; and $h = 0$.

45. The critical path has a length of 150 minutes.

47.

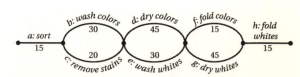

49. The critical path includes the tasks a, c, and f.

51. Tasks (a, b) and (c, d, e) take place at the same time. Task f must be done alone.